U.S. Department of
Transportation

Research and Innovative
Technology Administration

Volpe National Transportation
Systems Center

DOT-VNTSC-FAA-05-09
PB2003-101662

# A Sensitivity Analysis Comparing Two Procedures for Adjusting As-Measured Spectra to Reference Conditions

## SAE ARP 866A vs. ISO 9613-1/ANSI S1.26-1995

Final Report
April 2002

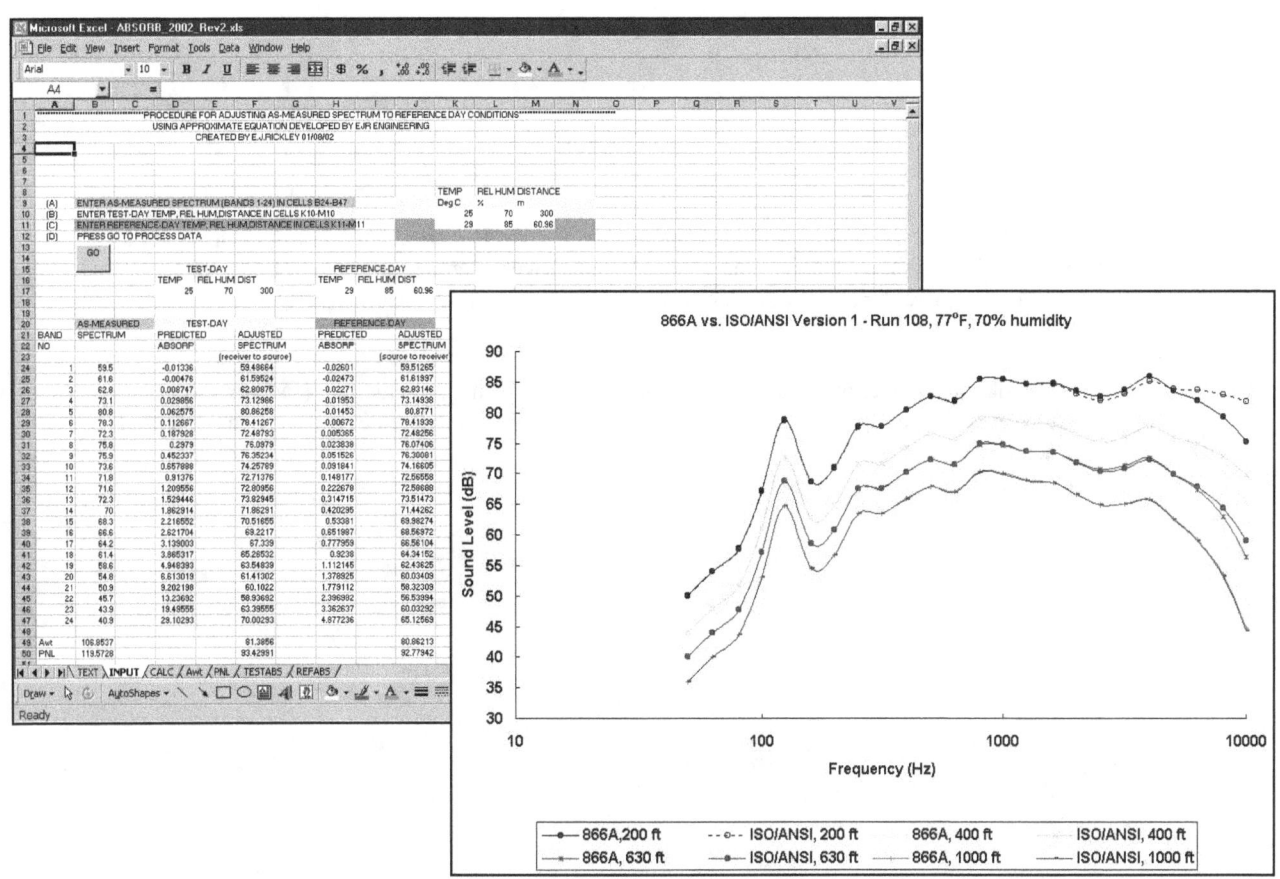

## NOTICE

This document is disseminated under the sponsorship of the Department of Transportation in the interest of information exchange. The United States Government assumes no liability for its contents or use thereof. This report does not constitute a standard, specification, or regulation.

The United States Government does not endorse products or manufacturers. Trade or manufacturers' names appear herein solely because they are considered essential to the object of this document.

| | | |
|---|---|---|
| # REPORT DOCUMENTATION PAGE | | *Form Approved*<br>*OMB No. 0704-0188* |

Public reporting burden for this collection of information is estimated to average 1 hour per response, including the time for reviewing instructions, searching existing data sources, gathering and maintaining the data needed, and completing and reviewing the collection of information. Send comments regarding this burden estimate or any other aspect of this collection of information, including suggestions for reducing this burden, to Washington Headquarters Services, Directorate for Information Operations and Reports, 1215 Jefferson Davis Highway, Suite 1204, Arlington, VA 22202-4302, and to the Office of Management and Budget, Paperwork Reduction Project (0704-0188), Washington, DC 20503.

| 1. AGENCY USE ONLY (Leave blank) | 2. REPORT DATE<br>April 2002 | 3. REPORT TYPE AND DATES COVERED<br>Final Report<br>January 2002 – April 2002 |
|---|---|---|
| 4. TITLE AND SUBTITLE<br>A Sensitivity Analysis Comparing Two Procedures for Adjusting As-Measured Spectra to Reference Conditions | | 5 FUNDING NUMBERS<br>FA-53/BS232 |
| 6 AUTHOR(S)<br>Reherman, Clay N. [1], Roof, Christopher J. [1], Fleming, Gregg G. [1], Rickley, Edward J. [2] | | |
| 7. PERFORMING ORGANIZATION NAME(S) AND ADDRESS(ES)<br>U.S. Department of Transportation<br>Research and Innovative Technology Administration<br>John A. Volpe National Transportation Systems Center<br>Environmental Measurement and Modeling Division, DTS-34<br>Cambridge, MA 02142 | | 8. PERFORMING ORGANIZATION REPORT NUMBER<br>DOT-VNTSC-FAA-05-09 |
| 9. SPONSORING/MONITORING AGENCY NAME(S) AND ADDRESS(ES)<br>U.S. Department of Transportation<br>Federal Aviation Administration<br>Office of Environment and Energy<br>Washington, DC 20591 | | 10. SPONSORING/MONITORING AGENCY REPORT NUMBER |
| 11. SUPPLEMENTARY NOTES<br>(1) U.S. Department of Transportation    (2) EJR Engineering<br>Research and Innovative Technology Administration  10 Lorenzo Circle<br>John A. Volpe National Transportation Systems Center  Methuen, MA 01844<br>Environmental Measurement and Modeling Division<br>Cambridge, MA 02142 | | |
| 12a. DISTRIBUTION/AVAILABILITY STATEMENT<br>This document is available to the public through the National Technical Information Service (NTIS), Springfield, VA 22161 (NTIS # PB2003-101662) | | 12b. DISTRIBUTION CODE |

13. ABSTRACT (Maximum 200 words)

The Society of Automotive Engineers' (SAE) Aerospace Recommended Practice (ARP) No. 866A (866A), and a procedure utilizing pure-tone absorption equations developed in support of the International Organization for Standardization's (ISO) 9613-1 and the American National Standards Institute's (ANSI) S1.26-1995, are compared and a sensitivity analysis undertaken. The ISO and ANSI equations are identical to each other. Thirteen takeoff and two approach spectra for aircraft were utilized in the analysis, as well as one takeoff, two approach, and three flyover spectra for helicopters. As-measured spectra were processed at 10 standard distances between 200 ft. and 25,000 ft., and at five different temperature/ relative humidity (RH) combinations, to generate adjusted spectra and associated $L_{ASmx}$ values. The focus of the study is a refined version of the ISO/ANSI adjustment procedure developed specifically for application to one-third octave-level data.

| 14. SUBJECT TERMS<br>Society of Automotive Engineers, International Organization for Standardization, noise levels, dBA, atmospheric absorption, one-third octave noise spectra | | | 15. NUMBER OF PAGES<br>75 |
|---|---|---|---|
| | | | 16. PRICE CODE |
| 17. SECURITY CLASSIFICATION OF REPORT<br>Unclassified | 18. SECURITY CLASSIFICATION OF THIS PAGE<br>Unclassified | 19. SECURITY CLASSIFICATION OF ABSTRACT<br>Unclassified | 20. LIMITATION OF ABSTRACT  Unlimited |

NSN 7540-01-280-5500

Standard Form 298(Rev. 2-89)
Prescribed by ANSI Std. 239-18 298-102

# METRIC/ENGLISH CONVERSION FACTORS

## ENGLISH TO METRIC | METRIC TO ENGLISH

### LENGTH (APPROXIMATE)

| ENGLISH TO METRIC | METRIC TO ENGLISH |
|---|---|
| 1 inch (in) = 2.5 centimeters (cm) | 1 millimeter (mm) = 0.04 inch (in) |
| 1 foot (ft) = 30 centimeters (cm) | 1 centimeter (cm) = 0.4 inch (in) |
| 1 yard (yd) = 0.9 meter (m) | 1 meter (m) = 3.3 feet (ft) |
| 1 mile (mi) = 1.6 kilometers (km) | 1 meter (m) = 1.1 yards (yd) |
|  | 1 kilometer (km) = 0.6 mile (mi) |

### AREA (APPROXIMATE)

| ENGLISH TO METRIC | METRIC TO ENGLISH |
|---|---|
| 1 square inch (sq in, in$^2$) = 6.5 square centimeters (cm$^2$) | 1 square centimeter (cm$^2$) = 0.16 square inch (sq in, in$^2$) |
| 1 square foot (sq ft, ft$^2$) = 0.09 square meter (m$^2$) | 1 square meter (m$^2$) = 1.2 square yards (sq yd, yd$^2$) |
| 1 square yard (sq yd, yd$^2$) = 0.8 square meter (m$^2$) | 1 square kilometer (km$^2$) = 0.4 square mile (sq mi, mi$^2$) |
| 1 square mile (sq mi, mi$^2$) = 2.6 square kilometers (km$^2$) | 10,000 square meters (m$^2$) = 1 hectare (ha) = 2.5 acres |
| 1 acre = 0.4 hectare (he) = 4,000 square meters (m$^2$) |  |

### MASS – WEIGHT (APPROXIMATE)

| ENGLISH TO METRIC | METRIC TO ENGLISH |
|---|---|
| 1 ounce (oz) = 28 grams (gm) | 1 gram (gm) = 0.036 ounce (oz) |
| 1 pound (lb) = 0.45 kilogram (kg) | 1 kilogram (kg) = 2.2 pounds (lb) |
| 1 short ton = 2,000 pounds (lb) = 0.9 tonne (t) | 1 tonne (t) = 1,000 kilograms (kg) = 1.1 short tons |

### VOLUME (APPROXIMATE)

| ENGLISH TO METRIC | METRIC TO ENGLISH |
|---|---|
| 1 teaspoon (tsp) = 5 milliliters (ml) | 1 milliliter (ml) = 0.03 fluid ounce (fl oz) |
| 1 tablespoon (tbsp) = 15 milliliters (ml) | 1 liter (l) = 2.1 pints (pt) |
| 1 fluid ounce (fl oz) = 30 milliliters (ml) | 1 liter (l) = 1.06 quarts (qt) |
| 1 cup © = 0.24 liter (l) | 1 liter (l) = 0.26 gallon (gal) |
| 1 pint (pt) = 0.47 liter (l) |  |
| 1 quart (qt) = 0.96 liter (l) |  |
| 1 gallon (gal) = 3.8 liters (l) |  |
| 1 cubic foot (cu ft, ft$^3$) = 0.03 cubic meter (m$^3$) | 1 cubic meter (m$^3$) = 36 cubic feet (cu ft, ft$^3$) |
| 1 cubic yard (cu yd, yd$^3$) = 0.76 cubic meter (m$^3$) | 1 cubic meter (m$^3$) = 1.3 cubic yards (cu yd, yd$^3$) |

### TEMPERATURE (EXACT)

| ENGLISH TO METRIC | METRIC TO ENGLISH |
|---|---|
| $[(x-32)(5/9)]\ °F = y\ °C$ | $[(9/5)y + 32]\ °C = x\ °F$ |

## QUICK INCH - CENTIMETER LENGTH CONVERSION

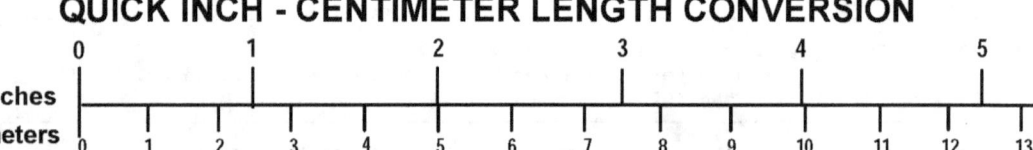

## QUICK FAHRENHEIT - CELSIUS TEMPERATURE CONVERSION

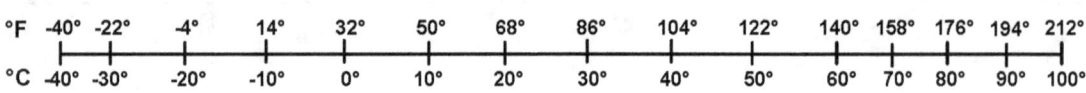

For more exact and or other conversion factors, see NIST Miscellaneous Publication 286, Units of Weights and Measures. Price $2.50 SD Catalog No. C13 10286.

# Table of Contents

| Section | Page |
|---|---|
| Table of Contents | i |
| List of Figures | ii |
| List of Tables | iii |
| Executive Summary | vi |
| 1 Introduction | 1 |
| 2 Analysis | 1 |
|    2.1 866A vs. the ISO/ANSI Spreadsheet | 4 |
|       2.1.1 Version 1, January 24, 2002 | 4 |
|       2.1.2 Version 2, February 11, 2002 | 7 |
| 3 Results | 10 |
| 4 Conclusions | 12 |
| Appendix A: Spectral Class Data | A-1 |
|    A.1 Modern Aircraft | A-1 |
|    A.2 Other Aircraft | A-2 |
|    A.3 Helicopters | A-6 |
| Appendix B: ISO/ANSI Pure-Tone Absorption Equation (Version 1 and Version 2) | B-1 |
|    B.1 Memorandum, January 31, 2002 | B-1 |
|       B.1.1 References | B-4 |
|    B.2 Addendum, March 15, 2002 | B-5 |
| Appendix C: $L_{ASmx}$ Difference Data Tables | C-1 |
|    C.1 As-Measured Spectra Adjusted to 77°F/70% RH | C-1 |
|    C.2 As-Measured Spectra Adjusted to 85°F/35% RH | C-8 |
|    C.3 As-Measured Spectra Adjusted to 85°F/85% RH | C-15 |
|    C.4 As-Measured Spectra Adjusted to 40°F/85% RH | C-23 |
|    C.5 As-Measured Spectra Adjusted to 40°F/55% RH | C-30 |

# List of Figures

| Figure | Page |
|---|---|
| ES-1. ISO/ANSI – 866A Difference Data Summary | vi |
| 1. The FAR 36 Allowable Test Window | 3 |
| 2. The ISO/ANSI Spreadsheet | 4 |
| 3. SC 101 $L_{ASmx}$ Data vs. Distance | 5 |
| 4. SC 108 $L_{ASmx}$ Data vs. Distance | 5 |
| 5. Spectral Data for SC 108 at Short Distances | 6 |
| 6. All Spectral Data for SC 108 | 7 |
| 7. All Spectral Data for SC 108 | 8 |
| 8. 85°F/35% $L_{ASmx}$ Data Using 866A and ISO/ANSI, Version 1 | 8 |
| 9. 85°F/35% $L_{ASmx}$ Data Using 866A and ISO/ANSI, Version 2 | 9 |
| 10. ISO/ANSI – 866A Difference Data Summary | 12 |
| A.1 Spectral Class 103 Sound Pressure Levels | A-1 |
| A.2 Spectral Class 105 Sound Pressure Levels | A-1 |
| A.3 Spectral Class 203 Sound Pressure Levels | A-2 |
| A.4 Spectral Class 205 Sound Pressure Levels | A-2 |
| A.5 Spectral Class 101 Sound Pressure Levels | A-2 |
| A.6 Spectral Class 102 Sound Pressure Levels | A-3 |
| A.7 Spectral Class 104 Sound Pressure Levels | A-3 |
| A.8 Spectral Class 106 Sound Pressure Levels | A-3 |
| A.9 Spectral Class 107 Sound Pressure Levels | A-4 |
| A.10 Spectral Class 108 Sound Pressure Levels | A-4 |
| A.11 Spectral Class 109 Sound Pressure Levels | A-4 |
| A.12 Spectral Class 110 Sound Pressure Levels | A-5 |
| A.13 Spectral Class 111 Sound Pressure Levels | A-5 |
| A.14 Spectral Class 112 Sound Pressure Levels | A-5 |
| A.15 Spectral Class 113 Sound Pressure Levels | A-6 |
| A.16 Spectral Class 116 Sound Pressure Levels | A-6 |
| A.17 Spectral Class 219 Sound Pressure Levels | A-6 |
| A.18 Spectral Class 222 Sound Pressure Levels | A-7 |
| A.19 Spectral Class 302 Sound Pressure Levels | A-7 |
| A.20 Spectral Class 307 Sound Pressure Levels | A-7 |
| A.21 MD900 Sound Pressure Levels | A-8 |
| B.1 Case 1--25°C/70%RH Exact Band-Level vs. Mid-Band-Level Attenuation Data. | B-5 |
| B.2 Case 2--25°C/70%RH Exact Band-Level vs. Mid-Band-Level Attenuation Data. | B-5 |

## List of Tables

| Table | | Page |
|---|---|---|
| 1. | Representative Modern Aircraft Spectral Classes. | 1 |
| 2. | Representative Other Aircraft Spectral Classes. | 2 |
| 3. | Helicopter Spectral Class Information. | 2 |
| 4. | $L_{ASmx}$ Difference Data for 866A vs. ISO/ANSI Version 2 at 77°F/70% RH. | 10 |
| 5. | $L_{ASmx}$ Difference Data for 866A vs. ISO/ANSI Version 2 at 85°F/35% RH. | 10 |
| 6. | $L_{ASmx}$ Difference Data for 866A vs. ISO/ANSI Version 2 at 85°F/85% RH. | 11 |
| 7. | $L_{ASmx}$ Difference Data for 866A vs. ISO/ANSI Version 2 at 40°F/85% RH. | 11 |
| 8. | $L_{ASmx}$ Difference Data for 866A vs. ISO/ANSI Version 2 at 40°F/55% RH. | 12 |
| C.1 | Spectral Class 101 at 77°F/70% RH. | C-1 |
| C.2 | Spectral Class 102 at 77°F/70% RH. | C-1 |
| C.3 | Spectral Class 103 at 77°F/70% RH. | C-2 |
| C.4 | Spectral Class 104 at 77°F/70% RH. | C-2 |
| C.5 | Spectral Class 105 at 77°F/70% RH. | C-2 |
| C.6 | Spectral Class 106 at 77°F/70% RH. | C-3 |
| C.7 | Spectral Class 107 at 77°F/70% RH. | C-3 |
| C.8 | Spectral Class 108 at 77°F/70% RH. | C-3 |
| C.9 | Spectral Class 109 at 77°F/70% RH. | C-4 |
| C.10 | Spectral Class 110 at 77°F/70% RH. | C-4 |
| C.11 | Spectral Class 111 at 77°F/70% RH. | C-4 |
| C.12 | Spectral Class 112 at 77°F/70% RH. | C-5 |
| C.13 | Spectral Class 113 at 77°F/70% RH. | C-5 |
| C.14 | Spectral Class 116 at 77°F/70% RH. | C-5 |
| C.15 | Spectral Class 203 at 77°F/70% RH. | C-6 |
| C.16 | Spectral Class 205 at 77°F/70% RH. | C-6 |
| C.17 | Spectral Class 219 at 77°F/70% RH. | C-6 |
| C.18 | Spectral Class 222 at 77°F/70% RH. | C-7 |
| C.19 | Spectral Class 302 at 77°F/70% RH. | C-7 |
| C.20 | Spectral Class 307 at 77°F/70% RH. | C-7 |
| C.21 | MD900 Spectral Data at 77°F/70% RH. | C-8 |
| C.22 | Spectral Class 101 at 85°F/35% RH. | C-8 |
| C.23 | Spectral Class 102 at 85°F/35% RH. | C-9 |
| C.24 | Spectral Class 103 at 85°F/35% RH. | C-9 |
| C.25 | Spectral Class 104 at 85°F/35% RH. | C-9 |
| C.26 | Spectral Class 105 at 85°F/35% RH. | C-10 |
| C.27 | Spectral Class 106 at 85°F/35% RH. | C-10 |
| C.28 | Spectral Class 107 at 85°F/35% RH. | C-10 |
| C.29 | Spectral Class 108 at 85°F/35% RH. | C-11 |
| C.30 | Spectral Class 109 at 85°F/35% RH. | C-11 |
| C.31 | Spectral Class 110 at 85°F/35% RH. | C-11 |
| C.32 | Spectral Class 111 at 85°F/35% RH. | C-12 |

## List of Tables (continued)

| Table | | Page |
|---|---|---|
| C.33 | Spectral Class 112 at 85°F/35% RH. | C-12 |
| C.34 | Spectral Class 113 at 85°F/35% RH. | C-12 |
| C.35 | Spectral Class 116 at 85°F/35% RH. | C-13 |
| C.36 | Spectral Class 203 at 85°F/35% RH. | C-13 |
| C.37 | Spectral Class 205 at 85°F/35% RH. | C-13 |
| C.38 | Spectral Class 219 at 85°F/35% RH. | C-14 |
| C.39 | Spectral Class 222 at 85°F/35% RH. | C-14 |
| C.40 | Spectral Class 302 at 85°F/35% RH. | C-14 |
| C.41 | Spectral Class 307 at 85°F/35% RH. | C-15 |
| C.42 | MD900 Spectral Data at 85°F/35% RH. | C-15 |
| C.43 | Spectral Class 101 at 85°F/85% RH. | C-16 |
| C.44 | Spectral Class 102 at 85°F/85% RH. | C-16 |
| C.45 | Spectral Class 103 at 85°F/85% RH. | C-16 |
| C.46 | Spectral Class 104 at 85°F/85% RH. | C-17 |
| C.47 | Spectral Class 105 at 85°F/85% RH. | C-17 |
| C.48 | Spectral Class 106 at 85°F/85% RH. | C-17 |
| C.49 | Spectral Class 107 at 85°F/85% RH. | C-18 |
| C.50 | Spectral Class 108 at 85°F/85% RH. | C-18 |
| C.51 | Spectral Class 109 at 85°F/85% RH. | C-18 |
| C.52 | Spectral Class 110 at 85°F/85% RH. | C-19 |
| C.53 | Spectral Class 111 at 85°F/85% RH. | C-19 |
| C.54 | Spectral Class 112 at 85°F/85% RH. | C-19 |
| C.55 | Spectral Class 113 at 85°F/85% RH. | C-20 |
| C.56 | Spectral Class 116 at 85°F/85% RH. | C-20 |
| C.57 | Spectral Class 203 at 85°F/85% RH. | C-20 |
| C.58 | Spectral Class 205 at 85°F/85% RH. | C-21 |
| C.59 | Spectral Class 219 at 85°F/85% RH. | C-21 |
| C.60 | Spectral Class 222 at 85°F/85% RH. | C-21 |
| C.61 | Spectral Class 302 at 85°F/85% RH. | C-22 |
| C.62 | Spectral Class 307 at 85°F/85% RH. | C-22 |
| C.63 | MD900 Spectral Data at 85°F/85% RH. | C-22 |
| C.64 | Spectral Class 101 at 40°F/85% RH. | C-23 |
| C.65 | Spectral Class 102 at 40°F/85% RH. | C-23 |
| C.66 | Spectral Class 103 at 40°F/85% RH. | C-24 |
| C.67 | Spectral Class 104 at 40°F/85% RH. | C-24 |
| C.68 | Spectral Class 105 at 40°F/85% RH. | C-24 |
| C.69 | Spectral Class 106 at 40°F/85% RH. | C-25 |
| C.70 | Spectral Class 107 at 40°F/85% RH. | C-25 |
| C.71 | Spectral Class 108 at 40°F/85% RH. | C-25 |
| C.72 | Spectral Class 109 at 40°F/85% RH. | C-26 |
| C.73 | Spectral Class 110 at 40°F/85% RH. | C-26 |
| C.74 | Spectral Class 111 at 40°F/85% RH. | C-26 |

## List of Tables (continued)

| Table | | Page |
|---|---|---|
| C.75 | Spectral Class 112 at 40°F/85% RH. | C-27 |
| C.76 | Spectral Class 113 at 40°F/85% RH. | C-27 |
| C.77 | Spectral Class 116 at 40°F/85% RH. | C-27 |
| C.78 | Spectral Class 203 at 40°F/85% RH. | C-28 |
| C.79 | Spectral Class 205 at 40°F/85% RH. | C-28 |
| C.80 | Spectral Class 219 at 40°F/85% RH. | C-28 |
| C.81 | Spectral Class 222 at 40°F/85% RH. | C-29 |
| C.82 | Spectral Class 302 at 40°F/85% RH. | C-29 |
| C.83 | Spectral Class 307 at 40°F/85% RH. | C-29 |
| C.84 | MD900 Spectral Data at 40°F/85% RH. | C-30 |
| C.85 | Spectral Class 101 at 40°F/55% RH. | C-30 |
| C.86 | Spectral Class 102 at 40°F/55% RH. | C-31 |
| C.87 | Spectral Class 103 at 40°F/55% RH. | C-31 |
| C.88 | Spectral Class 104 at 40°F/55% RH. | C-31 |
| C.89 | Spectral Class 105 at 40°F/55% RH. | C-32 |
| C.90 | Spectral Class 106 at 40°F/55% RH. | C-32 |
| C.91 | Spectral Class 107 at 40°F/55% RH. | C-32 |
| C.92 | Spectral Class 108 at 40°F/55% RH. | C-33 |
| C.93 | Spectral Class 109 at 40°F/55% RH. | C-33 |
| C.94 | Spectral Class 110 at 40°F/55% RH. | C-33 |
| C.95 | Spectral Class 111 at 40°F/55% RH. | C-34 |
| C.96 | Spectral Class 112 at 40°F/55% RH. | C-34 |
| C.97 | Spectral Class 113 at 40°F/55% RH. | C-34 |
| C.98 | Spectral Class 116 at 40°F/55% RH. | C-35 |
| C.99 | Spectral Class 203 at 40°F/55% RH. | C-35 |
| C.100 | Spectral Class 205 at 40°F/55% RH. | C-35 |
| C.101 | Spectral Class 219 at 40°F/55% RH. | C-36 |
| C.102 | Spectral Class 222 at 40°F/55% RH. | C-36 |
| C.103 | Spectral Class 302 at 40°F/55% RH. | C-36 |
| C.104 | Spectral Class 307 at 40°F/55% RH. | C-37 |
| C.105 | MD900 Spectral Data at 40°F/55% RH. | C-37 |

## Executive Summary

The Society of Automotive Engineers' (SAE) Aerospace Recommended Practice (ARP) No. 866A (866A), and a procedure utilizing pure-tone absorption equations developed in support of the International Organization for Standardization's (ISO) 9613-1 and the American National Standards Institute's (ANSI) S1.26-1995, are compared and a sensitivity analysis undertaken. The ISO and ANSI equations are identical to each other. Thirteen takeoff and two approach spectra for aircraft were utilized in the analysis, as well as one takeoff, two approach, and three flyover spectra for helicopters. As-measured spectra were processed at 10 standard distances between 200 ft. and 25,000 ft., and at five different temperature/ relative humidity (RH) combinations, to generate adjusted spectra and associated $L_{ASmx}$ values. The focus of the study is a refined version of the ISO/ANSI adjustment procedure developed specifically for application to one-third octave-level data.

The results of the study are summarized in Figure ES-1.

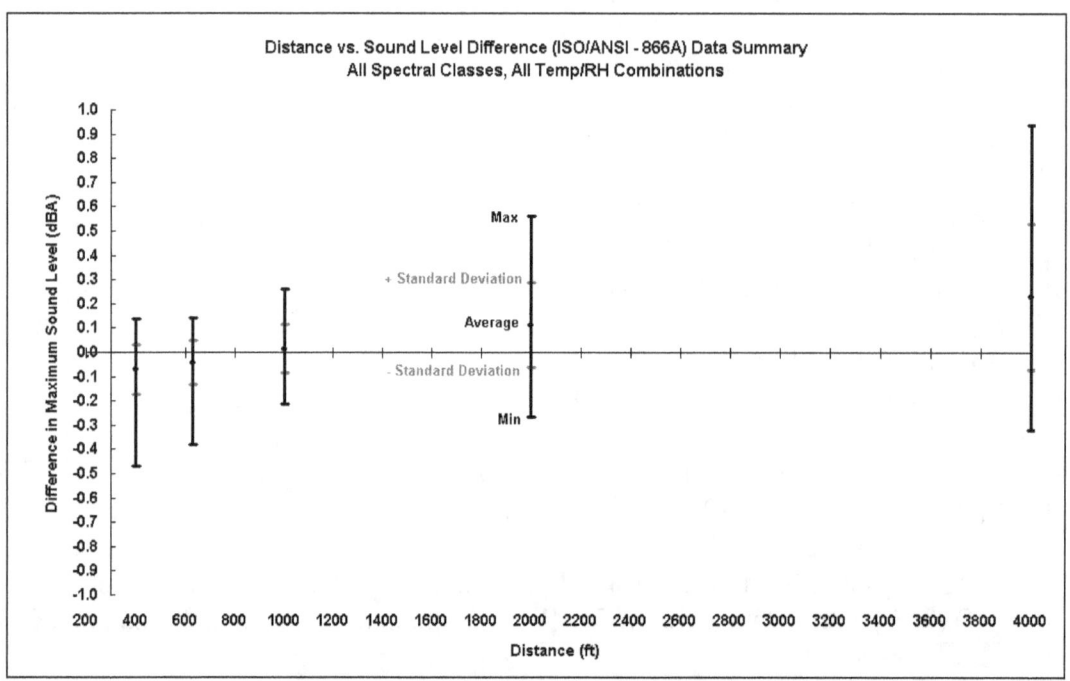

**Figure ES-1.** ISO/ANSI – 866A Difference Data Summary

Difference values were obtained by subtracting 866A results from ISO/ANSI results. For aircraft noise certification purposes, distances between 630 ft. and 2,000 ft. may be considered most relevant. Therefore, the average, range, and standard deviations for all spectral data and temperature/RH combinations at 400 ft., 630 ft., 1,000 ft., 2,000 ft., and 4,000 ft. are presented. At 400 ft., the maximum difference for all the data is 0.1 dBA, the minimum difference is –0.5 dBA, the average difference is –0.1 dBA, and the standard deviation is 0.1 dBA. At 630 ft., the maximum difference is 0.1 dBA, the minimum difference is –0.4 dBA, the average difference is 0.0 dBA, and the standard deviation is 0.1 dBA.  At 1,000 ft., the maximum difference is 0.3 dBA, the minimum difference is –0.2 dBA, the average difference is 0.0 dBA, and the standard deviation is 0.1 dBA.  At 2,000 ft., the maximum difference is 0.6 dBA, the minimum difference is –0.3 dBA, the average difference is 0.1 dBA, and the standard deviation is 0.2 dBA.  At 4,000 ft., the maximum difference is 0.9 dBA, the minimum difference is –0.3 dBA, the average difference is 0.2 dBA, and the standard deviation is 0.3 dBA.

Representative plots of the $L_{ASmx}$ data as well as A-weighted, spectra at time of $L_{ASmx}$ are presented.

# 1 Introduction

Two procedures for adjusting as-measured test-day spectra to reference day conditions – the Society of Automotive Engineers' (SAE) Aerospace Recommended Practice (ARP) No. 866A (866A) and a procedure utilizing pure-tone absorption equations, developed in support of the International Organization for Standardization's (ISO) 9613-1 and the American National Standards Institute's (ANSI) S1.26-1995, and refined for application to one-third octave-band data – were evaluated and compared. The ISO and ANSI pure-tone absorption equations are identical to each other. The 866A procedure historically has been used for atmospheric absorption corrections for Federal Aviation Regulations (FAR) Part 36, Noise Standards: Aircraft Type and Airworthiness Certification and similar regulations of the International Civil Aviation Organization (ICAO). The ISO/ANSI procedure analyzed herein is under consideration for inclusion into harmonized regulations. This document presents a comparison of data corrected to several distances using the two different methodologies.

# 2 Analysis

Twenty-one sets of spectral data were utilized in the analysis, including 14 airplane and helicopter takeoffs, four approaches, and two level flyovers from the Federal Aviation Administration's (FAA) Integrated Noise Model (INM), and an MD900 level flyover from a 1996 joint government/industry noise measurement at Crow's Landing, California, sponsored by the National Rotorcraft Technology Center (NRTC).

Table 1 lists spectral class (SC) data for several aircraft chosen because they are considered to be representative of *modern* jet aircraft.

**Table 1.** Representative Modern Aircraft Spectral Classes.

| SC | Flight Type | Description | Aircraft ID |
|---|---|---|---|
| 103 | Departure | Two engine high bypass ratio turbofan aircraft | 757PW / 757RR / 767300 / 767CF6 / 767JT9 / A300 / A310 / A320 |
| 105 | Departure | Two engine high bypass ratio turbofan aircraft | MD9025 / MD9028 / 777200 |
| 203 | Approach | Two engine high bypass ratio turbofan aircraft; two engine turbofan business aircraft | 757PW / 757RR / 767300 / 767CF6 / 767JT9 / A300 / A310 / A320 / DC1010 / DC1030 / DC1040 / L1011 / L10115 / FAL20 / SABR80 / MU3001 |
| 205 | Approach | Two engine high bypass ratio turbofan aircraft | MD9025 / MD9028 / 777200 |

Table 2 presents a list of the other jet aircraft (and representative SC's) utilized for this analysis.

**Table 2.** Representative Other Aircraft Spectral Classes.

| SC | Flight Type | Description | Aircraft ID |
|---|---|---|---|
| 101 | Departure | Two and Three engine low bypass ratio turbofan aircraft | 727100 / 727200 / 727D15 / 727D17 / 727EM1 / 727EM2 / 727Q15 / 727Q7 / 727Q9 / 727QF / 737 / 737D17 / 737QN / DC9Q9 / DC910 / DC930 / DC950 / DC9Q7 / F10062 / F10065 / GIV / DC1010 / DC1030 / DC1040 / L1011 / L10115 / DC93HW / DC93LW |
| 102 | Departure | Two engine high bypass ratio turbofan aircraft | 737300 / 7373B2 / 737400 / 737500 |
| 104 | Departure | Two engine high bypass ratio turbofan aircraft | MD81 / MD82 / MD83 / F28MK2 / F28MK4 / GIIB / 737N17 / 737N19 |
| 106 | Departure | Four engine turbofan aircraft and supersonic turbojet aircraft | DC870 / 707QN / DC8QN / CONCRD |
| 107 | Departure | Four engine turbofan aircraft | 74710Q / 747200 / 74720A / 74420B / 747400 / 747SP / 747100 / 707120 / 707320 / 720B / DC850 / DC860 / 707 / 720 / DC820 |
| 108 | Departure | Four engine turbofan aircraft | BAE146 / BAC111 / BAE300 |
| 109 | Departure | Two engine turboprop; one and two engine piston aircraft | DHC6 / DHC6QP / SD330 / JPATS / GASEPF / GASEPV / BEC58P / COMSEP |
| 110 | Departure | Two engine turboprop; two and four engine piston aircraft | SF340 / HS748A / DC3 / DC6 |
| 111 | Departure | Two engine turboprop aircraft | CNA441 / C12 / C23 / T44 / U21 |
| 112 | Departure | Four engine turboprop aircraft | L188 / CVR580 / DHC7 / DHC8 / DHC830 / C-130E / C130 / C130AD / C130E / C130HP |
| 113 | Departure | Two engine turbojet and turbofan business aircraft | LEAR25 / LEAR35 / FAL20 / SABR80 / CNA500 / MU3001 / IA1125 / CIT3 / CL600 / CL601 / EMB145 / COMJET |

Table 3 presents a list of the helicopters and their associated SC's, as utilized in this analysis.

**Table 3.** Helicopter Spectral Class Information.

| SC | Flight Type | Aircraft ID |
|---|---|---|
| 116 | Departure | SA355 / S65 / H500D |
| 219 | Approach | S76 / SA350 / SA341 / SA365 |
| 222 | Approach | B206L |
| 302 | Flyover | S76 / SA341 / SA365 |
| 307 | Flyover | B206L |

See Appendix A for the normalized sound pressure levels of the 20 INM SC and also the MD900 helicopter used in the analysis.

For both the 866A and ISO/ANSI procedures, aircraft spectra were first corrected to the source (i.e., 0 ft.) assuming test-day conditions of 77°F, 70% relative humidity (RH), and a test distance of 1,000 ft.* The aircraft spectra were then corrected from the source to 10 standard distances using the two procedures: 200 ft., 400 ft., 630 ft., 1,000 ft., 2,000 ft., 4,000 ft., 6,300 ft., 10,000 ft., 16,000 ft., and 25,000 ft. The results for each procedure were then compared.

Five different temperature/RH combinations, 77°F/70% RH, 85°F/35% RH, 85°F/85% RH, 40°F/85% RH, and 40°F/55% RH, were used for corrections to the 10 standard distances. As illustrated in Figure 1, the temperature/RH combinations fall within the FAR 36 allowable test window, yet span a majority of the anticipated values for correction.

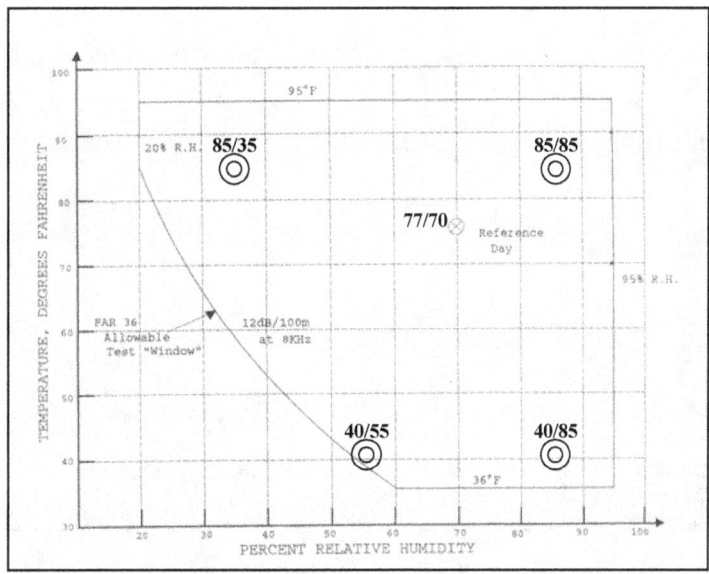

**Figure 1.** The FAR 36 Allowable Test Window

An internal Volpe computer program entitled LCorrect was used to implement the 866A procedure. This program accepts test-day temperature, humidity, and distance, reference temperature and humidity, and as-measured spectra as its inputs and calculates the spectrum at the time of A-weighted maximum sound level ($L_{ASmx}$) and the $L_{ASmx}$ for each of the 10 standard distances. LCorrect accepts speed and sound exposure level (SEL) inputs as well, but no speed-based corrections were applied and SEL is not used for the calculations documented herein.

A macro-driven Microsoft Excel spreadsheet developed by EJR Engineering was utilized to implement the ISO/ANSI procedure. The spreadsheet accepts test-day temperature, humidity, and distance, and reference temperature, humidity, distance, and spectra as

---

*Note that INM spectral class data are actually corrected to the SAE AIR 1845 Standard Atmosphere, not 77°F, 70% RH. The effects associated with this difference are consistent for the two procedures and therefore have no impact on the results.

inputs. It utilizes macros to calculate the $L_{ASmx}$ and the spectrum at time of $L_{ASmx}$ for a user-specified distance. Calculations are implemented using an empirically-derived regression equation (see Appendix B for a detailed description). An example of the user-interface page for the spreadsheet is presented in Figure 2.

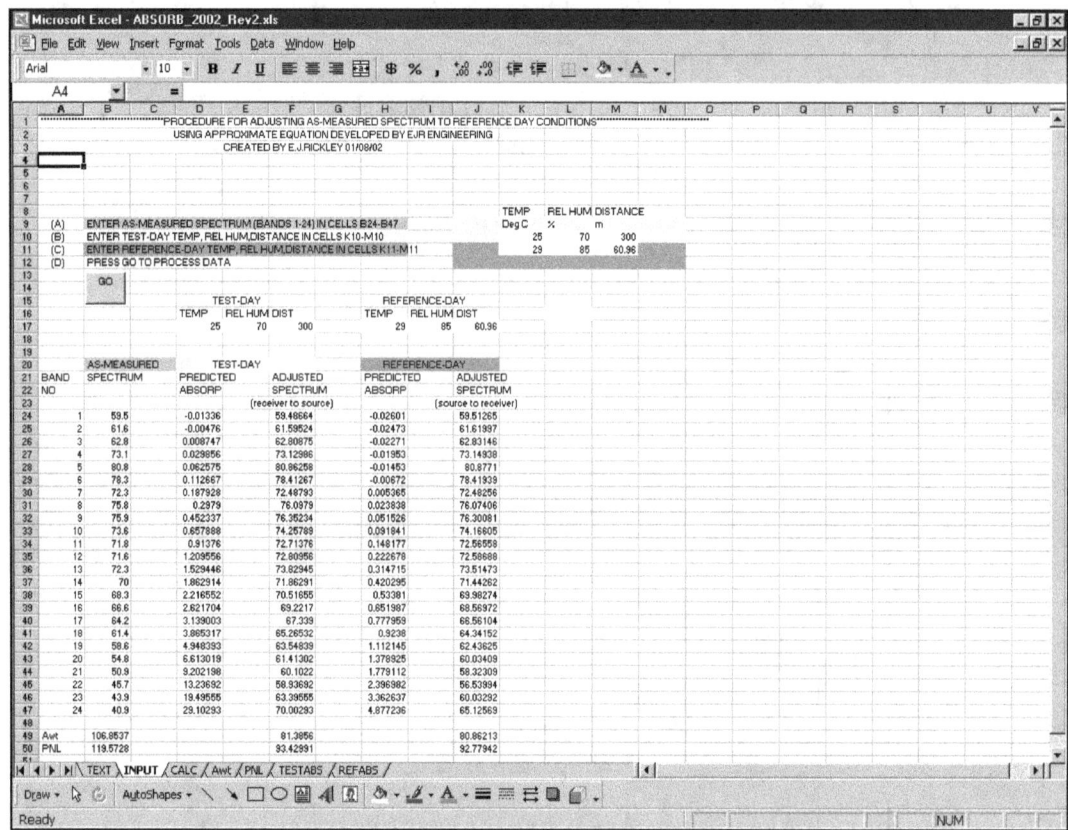

**Figure 2.** The ISO/ANSI Spreadsheet

The ANSI/ISO procedure must be repeated separately for corrections to multiple distances.

## 2.1 866A vs. the ISO/ANSI Spreadsheet

An initial version of the ISO/ANSI spreadsheet was tested and the results compared to 866A. Some anomalous results were observed using the first version; consequently, a second, improved version was developed and tested. The results of these comparisons are presented below in Section 2.1.1 for Version 1 of the spreadsheet and Section 2.1.2 for Version 2.

### 2.1.1 Version 1, January 24, 2002

$L_{ASmx}$ data were plotted as a function of distance for comparison. Figure 3 shows the results for SC 101, which consists of two- and three-engine low bypass ratio turbofan aircraft, including the 727-100 and 727-200, using 77°F/70% RH. The differences in dBA between the two procedures are shown using a secondary y-axis on the same plot.

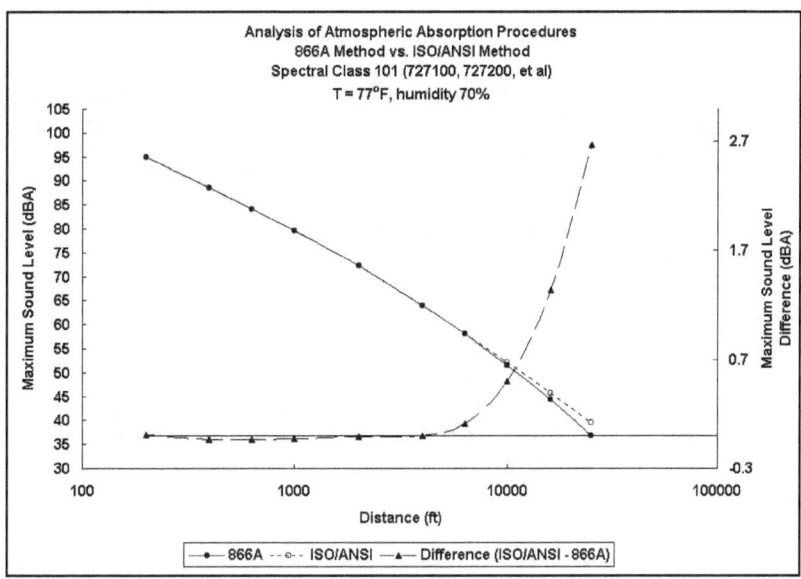

**Figure 3.** SC 101 $L_{ASmx}$ Data vs. Distance

As seen in Figure 3, corrections to 77°F/70% RH, 866A and the first version of the ISO/ANSI spreadsheet generally yielded similar results for distances between 200 ft. and 10,000 ft. (within 1.0 dBA). One of the largest ISO/ANSI – 866A differences found at a shorter distance was 0.2 dBA found at 200 ft. using SC 108 corrected to 77°F/70% RH. The results of this comparison are shown in Figure 4.

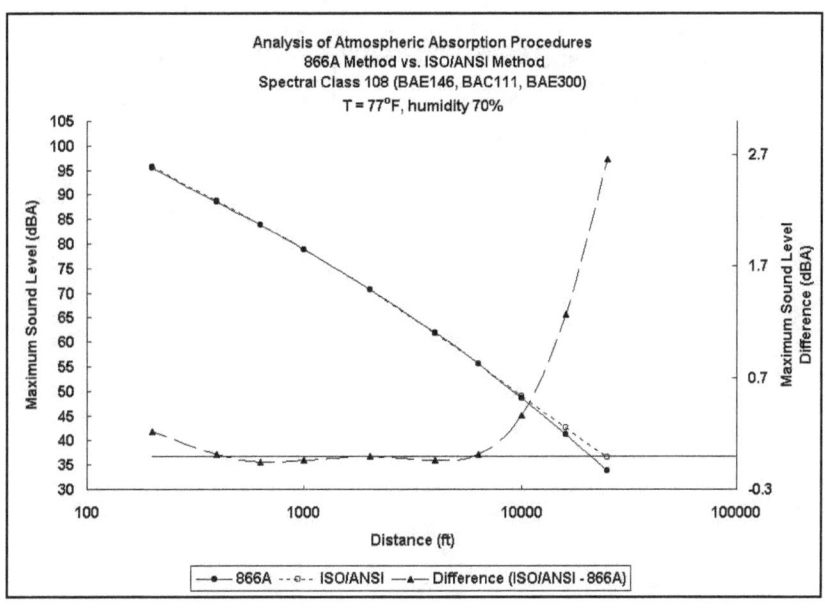

**Figure 4.** SC 108 $L_{ASmx}$ Data vs. Distance

The difference was 2.7 dBA at 25,000 ft. for SC 108. For distances between 16,000 ft. and 25,000 ft., the ISO/ANSI procedure consistently produces larger values, up to 2.9 dBA larger than 866A at 25,000 ft.

In Figure 5, an examination of the A-weighted spectra generated for SC 108 at 200, 400, 630, and 1,000 ft. reveals larger high-frequency energy for the ISO/ANSI procedure at 200 ft. In particular, the 6.5 dB difference in levels for 10,000 Hz one-third octave-band helps to explain the 0.2 dBA $L_{ASmx}$ difference at 200 ft. exhibited in Figure 4.

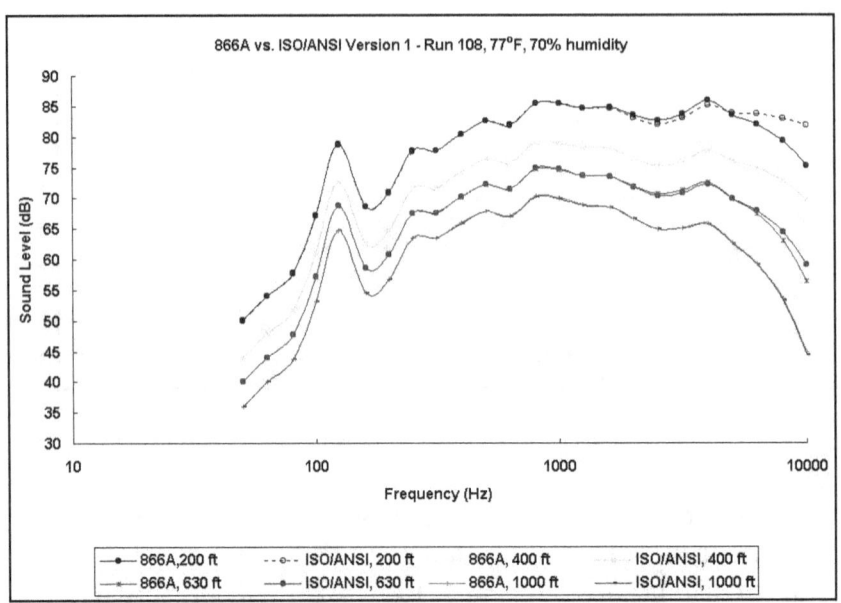

**Figure 5.** Spectral Data for SC 108 at Short Distances Generated Using 866A and ISO/ANSI Version 1

Figure 6 presents the A-weighted spectra for SC 108 plotted at each of the 10 standard distances. As can be seen, the first version of the ISO/ANSI spreadsheet tends to break down at higher frequencies for distances greater than 6,300 ft.

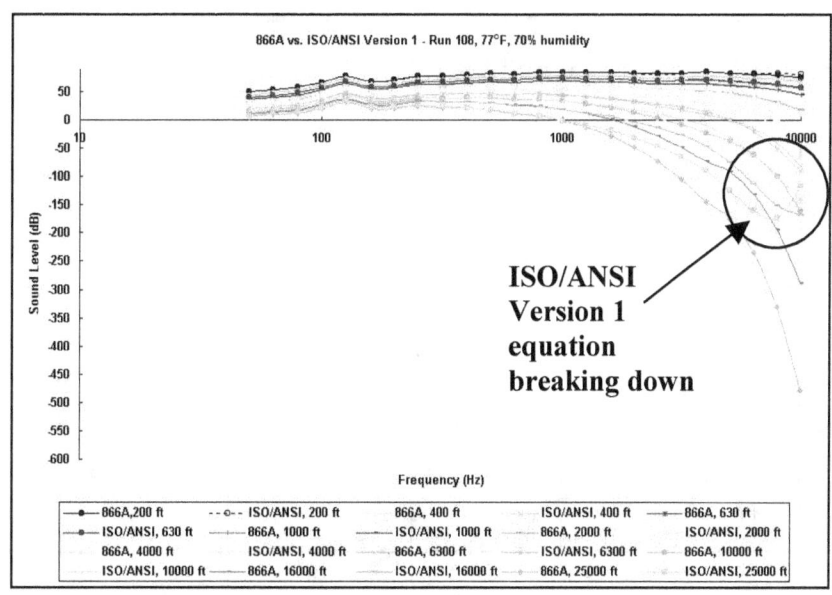

**Figure 6.** All Spectral Data for SC 108
Generated Using 866A and ISO/ANSI Version 1

Version 1 tests at 85°F/35% RH yield differences of up to 1.4 dBA for distances between 200 ft. and 16,000 ft., and differences of up to 10.5 dBA at 25,000 ft. Tests at 85°F/85% RH yield differences of up to 1.1 dBA for distances between 200 ft. and 6,300 ft., and differences of up to 4.9 dBA for distances between 10,000 ft. and 25,000 ft.

Tests were not completed for 40°F/85% RH and 40°F/55% RH because Version 1 of the ISO/ANSI spreadsheet would not calculate a 25,000-foot result for temperatures less than 54°F.

### 2.1.2 Version 2, February 11, 2002

Version 1 of the ISO/ANSI procedure was refined by EJR Engineering based on the results presented above, and Version 2 of the spreadsheet was developed. Tables in Appendix C present all of the $L_{ASmx}$ versus distance data generated using 866A and Version 2 of the ISO/ANSI spreadsheet.

$L_{ASmx}$ comparisons using Version 2 of the spreadsheet for 77°F/70% RH and 85°F/85% RH were very close to the results using the first version. As seen in Figure 7, plots of the A-weighted spectra generated by Version 2 illustrate more consistent high-frequency stability at larger distances. These improvements in the spectral data result in changes in the $L_{ASmx}$ values of less than 0.1 dBA, as compared to Version 1 of the ISO/ANSI spreadsheet.

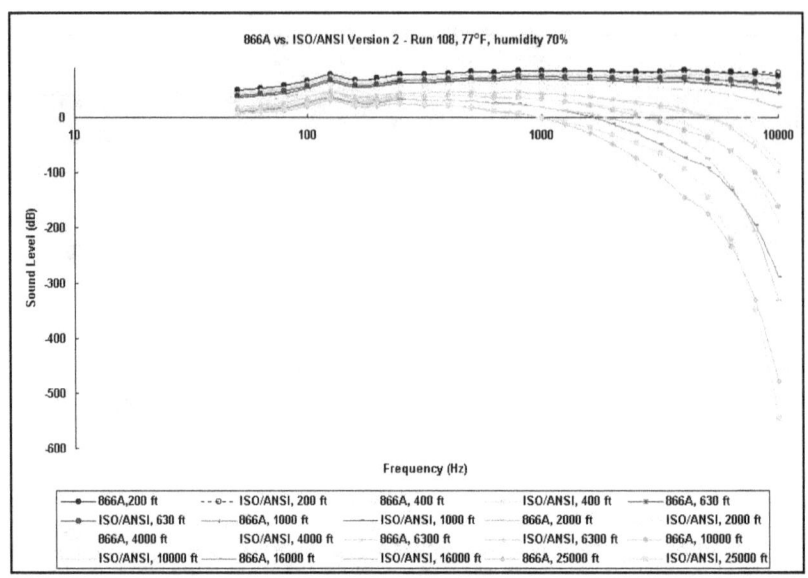

**Figure 7.** All Spectral Data for SC 108, Generated Using 866A and ISO/ANSI Version 2

For corrections to 85°F/35% RH, Version 2 of the ISO/ANSI spreadsheet yields results within 1.9 dBA of those calculated by 866A at 25,000 ft. For comparison, Figure 8 presents the results for the correction of SC 108 to 85°F/35% RH using the ISO/ANSI spreadsheet Version 1, whereas Figure 9 presents these same, improved results, calculated using ISO/ANSI Version 2.

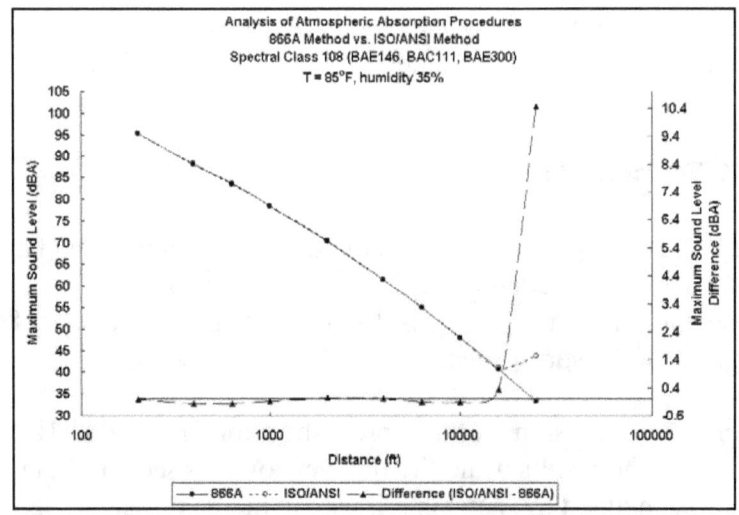

**Figure 8.** 85°F/35% $L_{ASmx}$ Data Using 866A and ISO/ANSI, Version 1

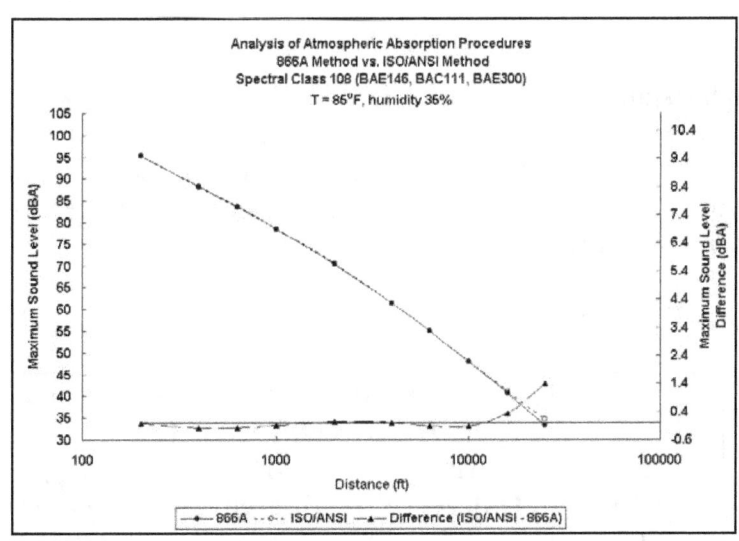

Figure 9. 85°F/35% $L_{ASmx}$ Data Using 866A and ISO/ANSI, Version 2

The ISO/ANSI spreadsheet Version 1 did not calculate sound levels at 25,000 feet for temperatures less than 54°F, but Version 2 does calculate these values. Tests at 40°F/85% RH yield differences of up to 1.8 dBA. Tests at 40°F/55% RH yield differences of up to 1.6 dBA.

Hereafter, only results computed using the second version of the ISO/ANSI spreadsheet will be presented.

## 3 Results

Tables 4 through 8 summarize the comparison of 866A $L_{ASmx}$ data with ISO/ANSI Version 2 $L_{ASmx}$ data for the five different temperature/RH combinations. These tables take into account all 21 spectral studies and present the maximum, minimum, and average difference values found at each distance. Difference data represent ISO/ANSI $L_{ASmx}$ values minus 866A $L_{ASmx}$ values.

**Table 4.** $L_{ASmx}$ Difference Data for 866A vs. ISO/ANSI Version 2 at 77°F/70% RH.

| Distance (ft.) | Difference Data: ISO/ANSI Version 2 $L_{ASmx}$ - 866A $L_{ASmx}$ (dBA) | | |
|---|---|---|---|
| | Max Difference | Min Difference | Average Difference |
| 200 | 0.2 | -0.3 | -0.1 |
| 400 | 0.0 | -0.2 | -0.1 |
| 630 | 0.0 | -0.2 | -0.1 |
| 1,000 | 0.0 | 0.0 | 0.0 |
| 2,000 | 0.1 | -0.1 | 0.0 |
| 4,000 | 0.3 | -0.2 | 0.0 |
| 6,300 | 0.5 | -0.2 | 0.1 |
| 10,000 | 1.0 | -0.1 | 0.4 |
| 16,000 | 1.8 | 0.7 | 1.2 |
| 25,000 | 2.9 | 2.2 | 2.5 |

**Table 5.** $L_{ASmx}$ Difference Data for 866A vs. ISO/ANSI Version 2 at 85°F/35% RH.

| Distance (ft.) | Difference Data: ISO/ANSI Version 2 $L_{ASmx}$ - 866A $L_{ASmx}$ (dBA) | | |
|---|---|---|---|
| | Max Difference | Min Difference | Average Difference |
| 200 | 0.0 | -0.4 | -0.1 |
| 400 | 0.0 | -0.4 | -0.1 |
| 630 | 0.0 | -0.3 | -0.1 |
| 1,000 | 0.0 | -0.2 | -0.1 |
| 2,000 | 0.1 | -0.2 | 0.0 |
| 4,000 | 0.1 | -0.3 | -0.1 |
| 6,300 | 0.3 | -0.4 | -0.2 |
| 10,000 | 0.6 | -0.6 | -0.3 |
| 16,000 | 1.1 | -0.5 | 0.0 |
| 25,000 | 1.9 | 0.1 | 0.8 |

**Table 6.** $L_{ASmx}$ Difference Data for 866A vs. ISO/ANSI Version 2 at 85°F/85% RH.

| Distance (ft.) | Difference Data: ISO/ANSI Version 2 $L_{ASmx}$ - 866A $L_{ASmx}$ (dBA) | | |
|---|---|---|---|
| | Max Difference | Min Difference | Average Difference |
| 200 | 0.4 | -0.3 | 0.0 |
| 400 | 0.1 | -0.3 | -0.1 |
| 630 | 0.1 | -0.2 | -0.1 |
| 1,000 | 0.1 | -0.2 | 0.0 |
| 2,000 | 0.3 | -0.3 | 0.0 |
| 4,000 | 0.6 | -0.2 | 0.3 |
| 6,300 | 1.1 | 0.2 | 0.7 |
| 10,000 | 1.8 | 1.0 | 1.5 |
| 16,000 | 3.1 | 2.4 | 2.8 |
| 25,000 | 4.9 | 4.1 | 4.7 |

**Table 7.** $L_{ASmx}$ Difference Data for 866A vs. ISO/ANSI Version 2 at 40°F/85% RH.

| Distance (ft.) | Difference Data: ISO/ANSI Version 2 $L_{ASmx}$ - 866A $L_{ASmx}$ (dBA) | | |
|---|---|---|---|
| | Max Difference | Min Difference | Average Difference |
| 200 | 0.0 | -0.5 | -0.1 |
| 400 | 0.0 | -0.5 | -0.1 |
| 630 | 0.1 | -0.4 | 0.0 |
| 1,000 | 0.2 | -0.2 | 0.0 |
| 2,000 | 0.4 | 0.0 | 0.2 |
| 4,000 | 0.7 | 0.1 | 0.4 |
| 6,300 | 1.0 | 0.1 | 0.6 |
| 10,000 | 1.3 | 0.2 | 0.8 |
| 16,000 | 1.5 | 0.4 | 1.0 |
| 25,000 | 1.8 | 0.6 | 1.1 |

**Table 8.** $L_{ASmx}$ Difference Data for 866A vs. ISO/ANSI Version 2 at 40°F/55% RH.

| Distance (ft.) | Difference Data: ISO/ANSI Version 2 $L_{ASmx}$ - 866A $L_{ASmx}$ (dBA) | | |
|---|---|---|---|
| | Max Difference | Min Difference | Average Difference |
| 200 | 0.0 | -0.5 | -0.1 |
| 400 | 0.1 | -0.4 | 0.0 |
| 630 | 0.1 | -0.2 | 0.0 |
| 1,000 | 0.3 | 0.0 | 0.1 |
| 2,000 | 0.6 | 0.0 | 0.3 |
| 4,000 | 0.9 | 0.1 | 0.5 |
| 6,300 | 1.1 | 0.1 | 0.7 |
| 10,000 | 1.3 | 0.1 | 0.8 |
| 16,000 | 1.6 | 0.2 | 0.9 |
| 25,000 | 1.6 | 0.2 | 1.0 |

## 4 Conclusions

Comparisons were made of 866A and ISO/ANSI $L_{ASmx}$ data for a range of distances spanning 200 to 25,000 feet. For certification purposes, distances between 630 and 2,000 feet may be considered most relevant. At these distances, differences between 866A and ISO/ANSI $L_{ASmx}$ data were practically negligible (±0.6 dBA). Difference value averages, ranges, and standard deviations for all SC and temperature/RH combinations at 400, 630, 1,000, 2,000, and 4,000 feet are presented in Figure 10.

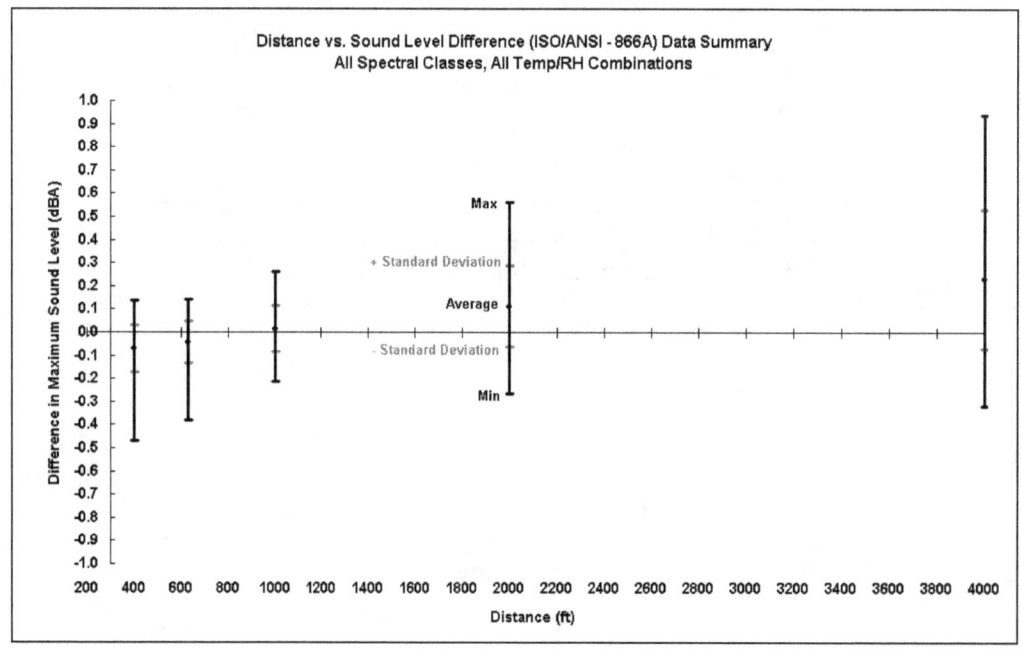

**Figure 10.** ISO/ANSI – 866A Difference Data Summary

As can be seen in Figure 10, the average, minimum and maximum differences for all data corrected to 400 feet are -0.1, -0.5 and 0.1 dBA, respectively. The standard deviation for the 400-foot data is 0.1 dBA. At 630 feet, the average, minimum and maximum differences are 0.0, -0.4 and 0.1 dBA, respectively. The standard deviation for the 630-foot data is 0.1 dBA. At 1,000 feet, the average, minimum and maximum differences are 0.0, -0.2 and 0.3 dBA, respectively. The standard deviation for the 100-foot data is 0.1 dBA. At 2,000 feet, the average, minimum and maximum differences are 0.1, -0.3 and 0.6 dBA, respectively. The 2,000-foot standard deviation is 0.2 dBA. At 4,000 feet, the average, minimum and maximum differences are 0.2, -0.3 and 0.9 dBA, respectively. The 4,000-foot standard deviation is 0.3 dBA.

# Appendix A: Spectral Class Data

The 20 sets of INM spectral class one-third octave-band data used in this study are presented below, along with the spectral data for the MD900. These data sets have been normalized in level, with relative differences in adjacent bands preserved; they are not actual data sets for these aircraft and helicopters.

## A.1 Modern Aircraft

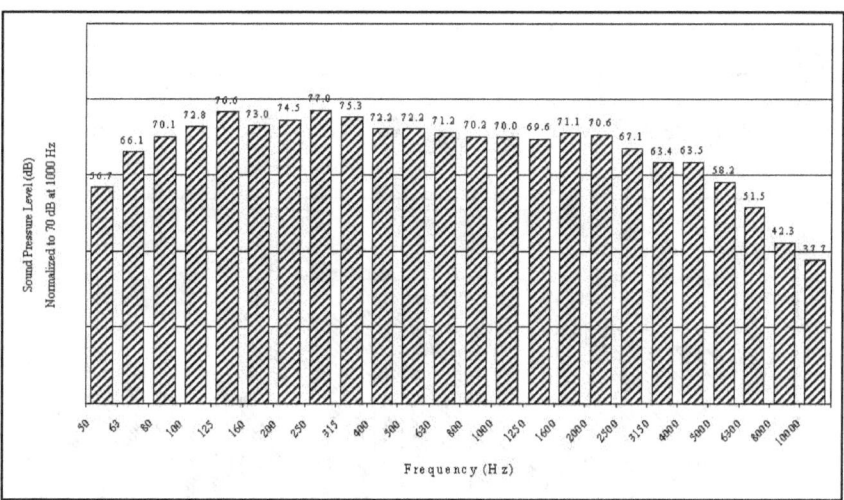

**Figure A.1** Spectral Class 103 Sound Pressure Levels

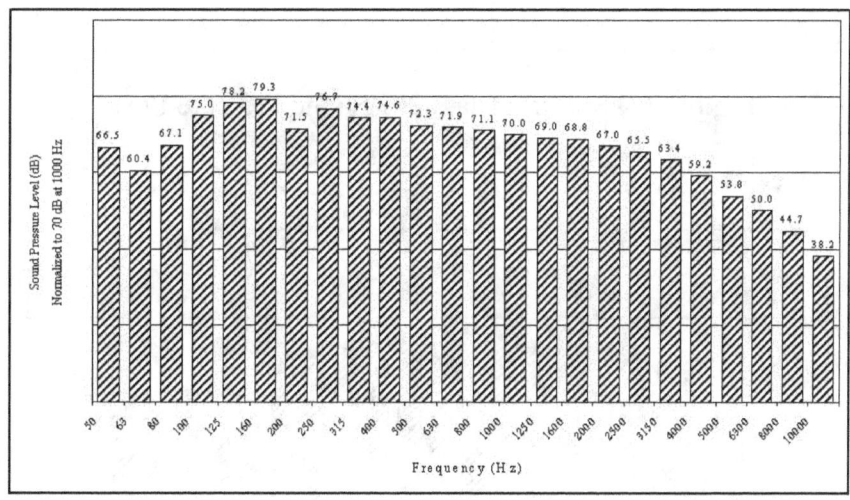

**Figure A.2** Spectral Class 105 Sound Pressure Levels

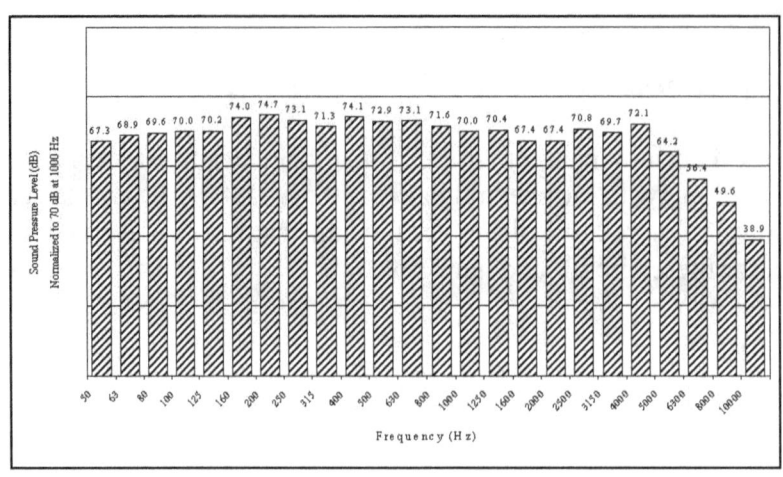

**Figure A.3** Spectral Class 203 Sound Pressure Levels

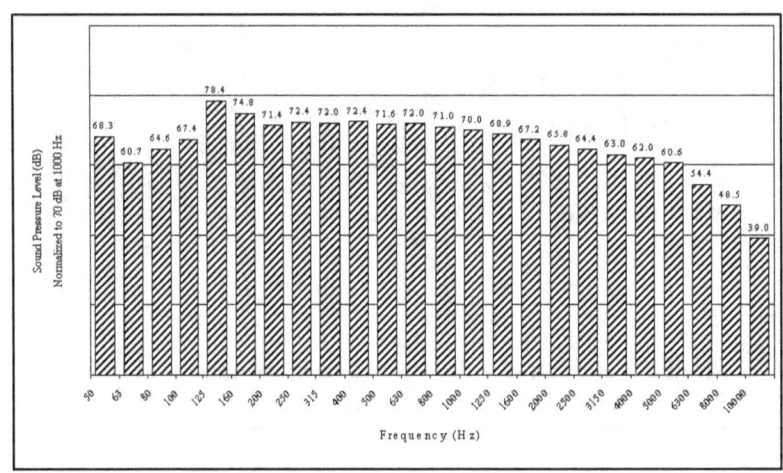

**Figure A.4** Spectral Class 205 Sound Pressure Levels

## A.2 Other Aircraft

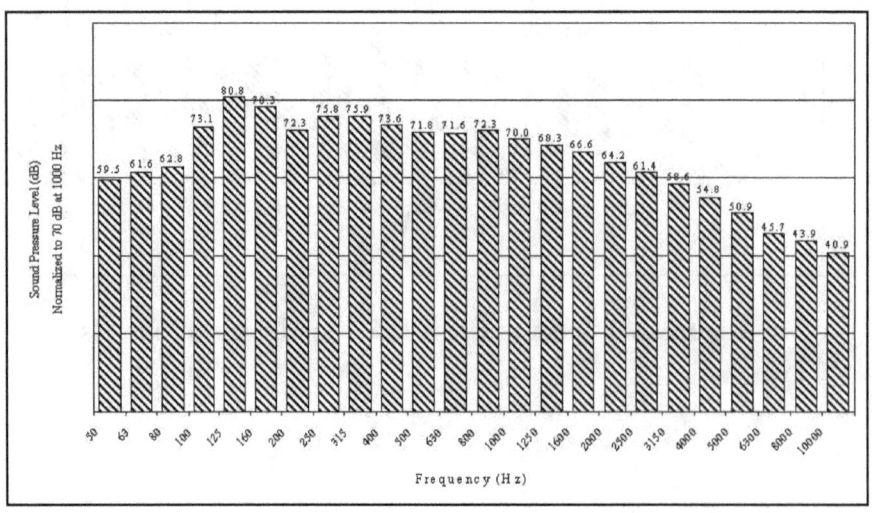

**Figure A.5** Spectral Class 101 Sound Pressure Levels

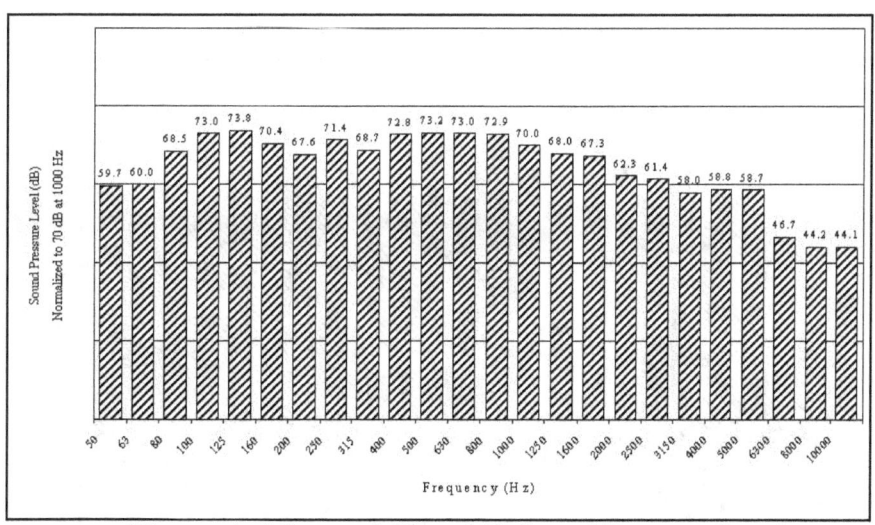

**Figure A.6** Spectral Class 102 Sound Pressure Levels

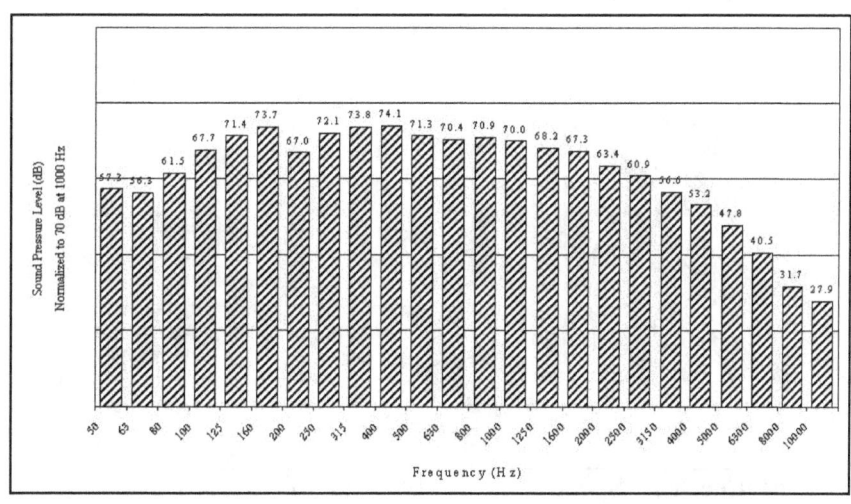

**Figure A.7** Spectral Class 104 Sound Pressure Levels

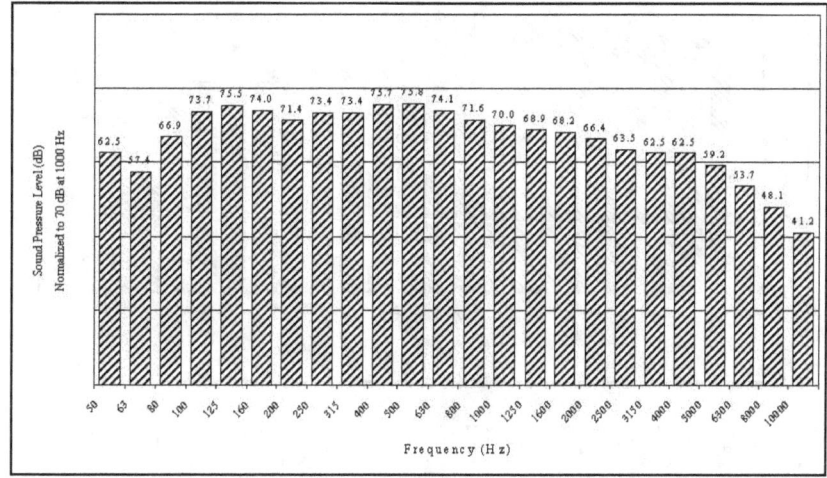

**Figure A.8** Spectral Class 106 Sound Pressure Levels

A-3

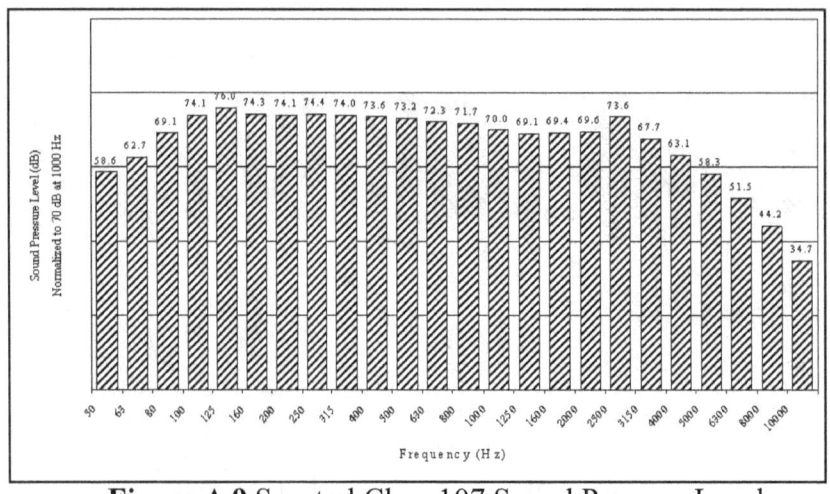

**Figure A.9** Spectral Class 107 Sound Pressure Levels

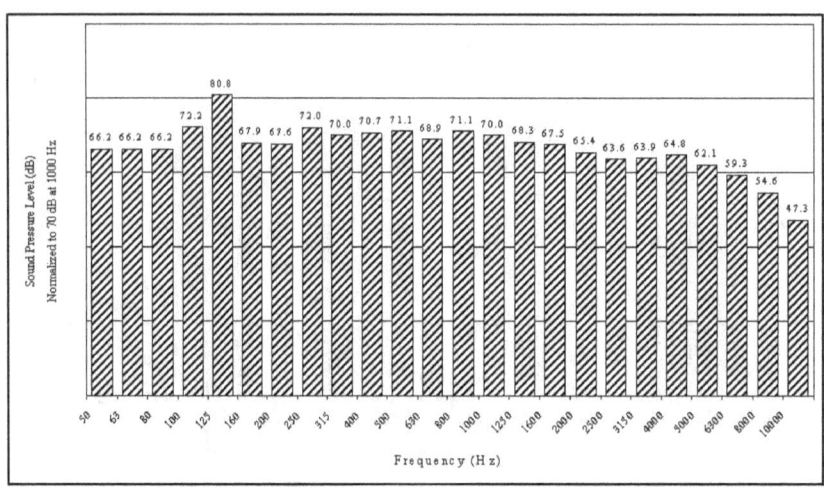

**Figure A.10** Spectral Class 108 Sound Pressure Levels

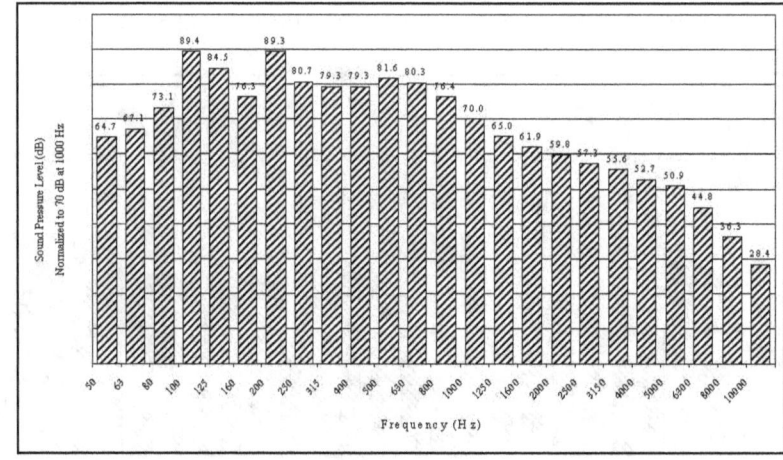

**Figure A.11** Spectral Class 109 Sound Pressure Levels

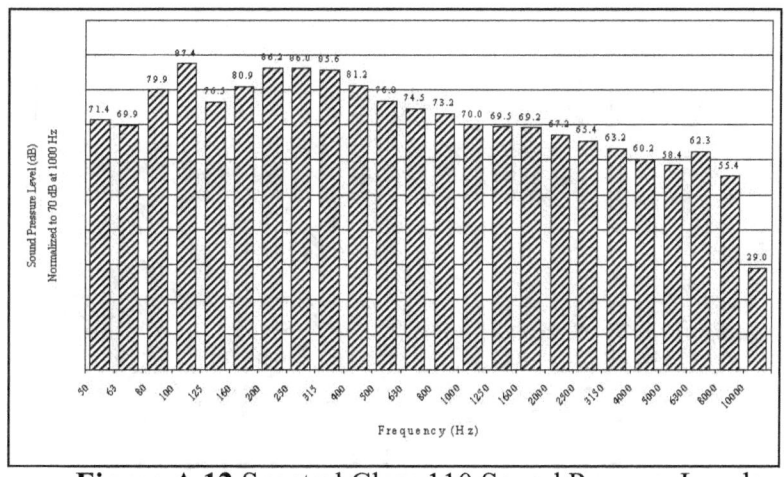

**Figure A.12** Spectral Class 110 Sound Pressure Levels

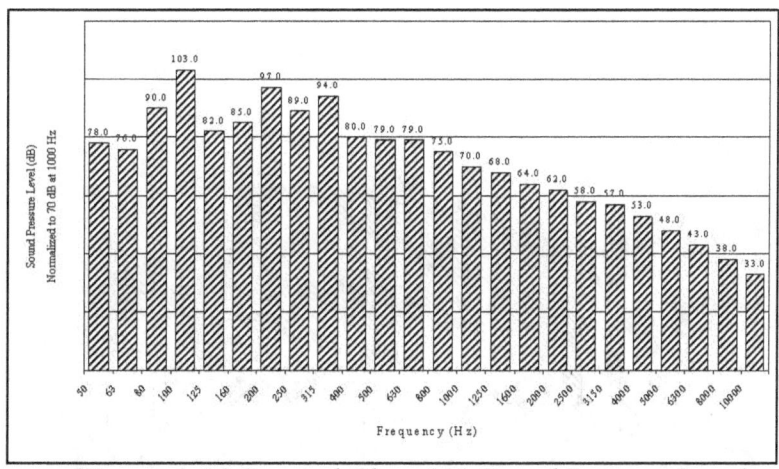

**Figure A.13** Spectral Class 111 Sound Pressure Levels

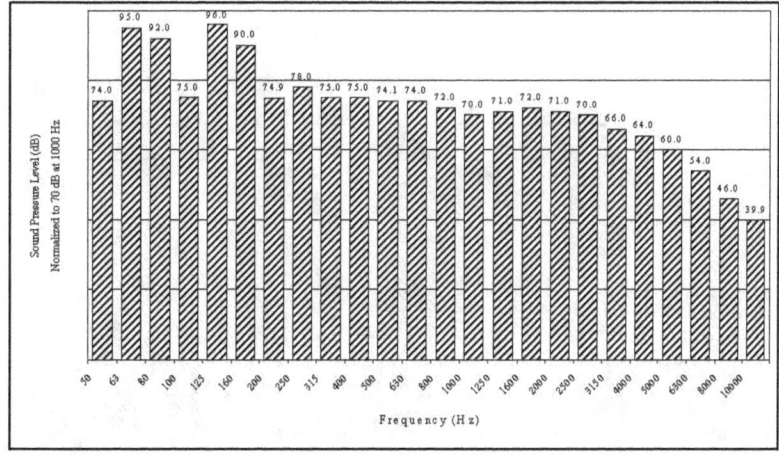

**Figure A.14** Spectral Class 112 Sound Pressure Levels

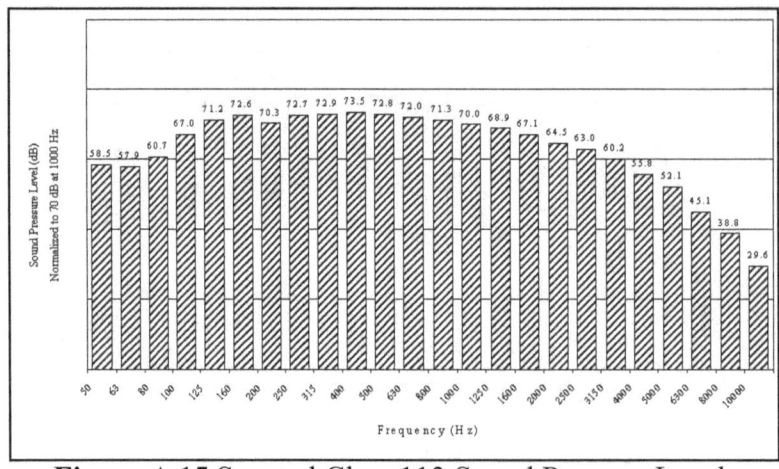

**Figure A.15** Spectral Class 113 Sound Pressure Levels

## A.3 Helicopters

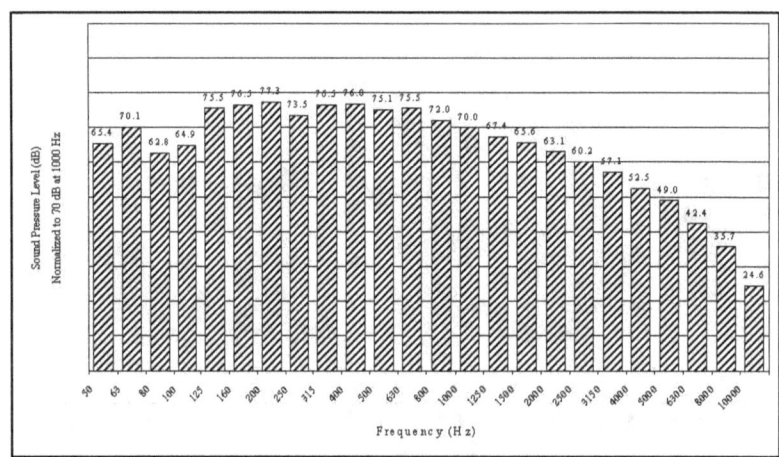

**Figure A.16** Spectral Class 116 Sound Pressure Levels

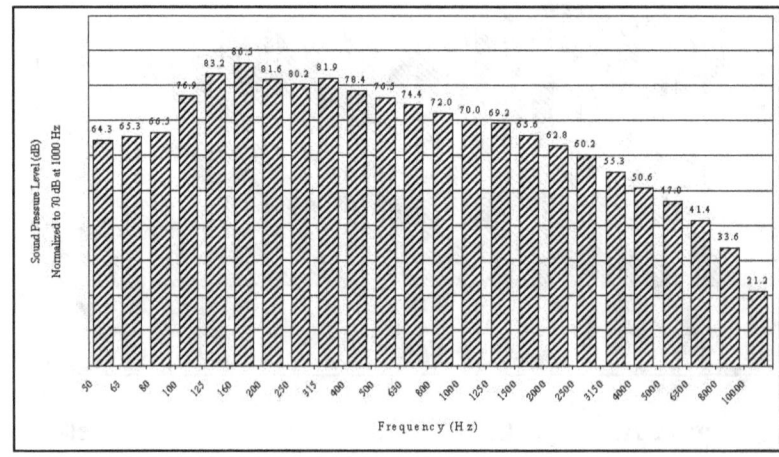

**Figure A.17** Spectral Class 219 Sound Pressure Levels

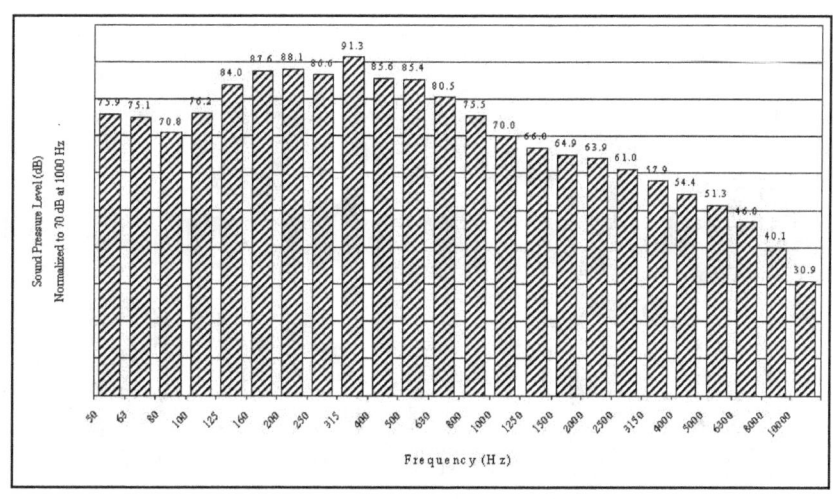

**Figure A.18** Spectral Class 222 Sound Pressure Levels

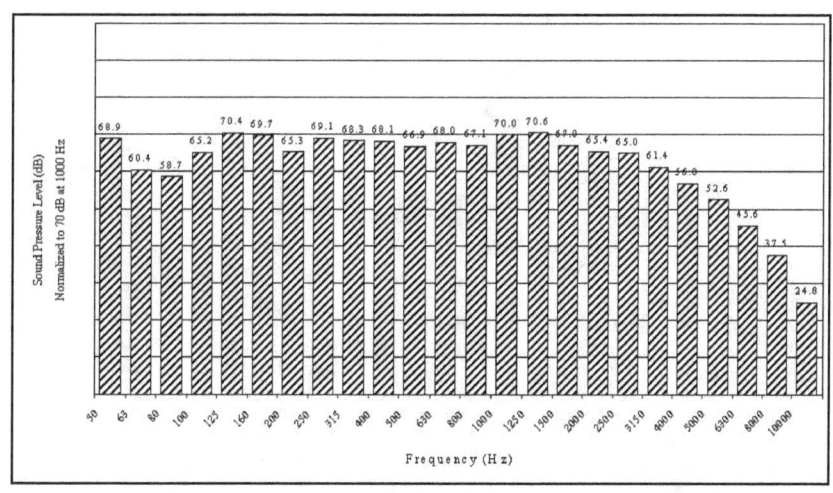

**Figure A.19** Spectral Class 302 Sound Pressure Levels

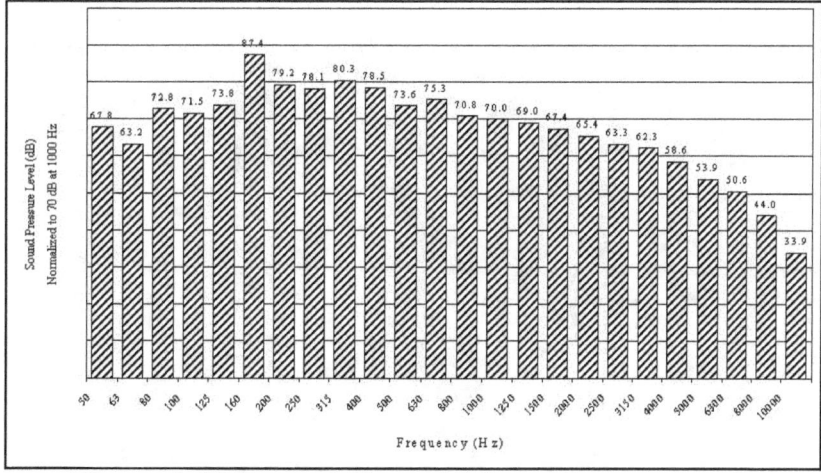

**Figure A.20** Spectral Class 307 Sound Pressure Levels

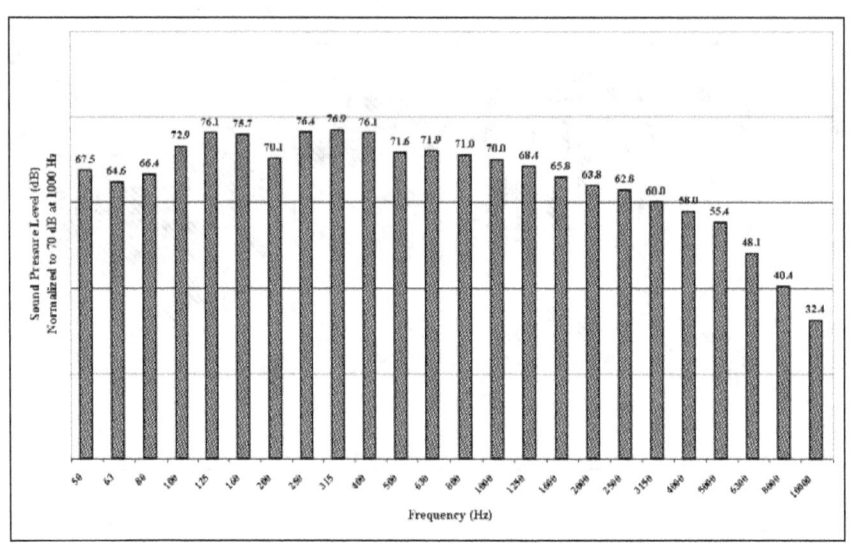

**Figure A.21** MD900 Sound Pressure Levels

**Appendix B: ISO/ANSI Pure-Tone Absorption Equation (Version 1 and Version 2)**

The following memorandum briefly explains the first version of the ISO/ANSI pure-tone absorption equation. An addendum has been added which briefly explains the changes made to the first version of the equation which led to the second version.

**B.1 Memorandum, January 31, 2002**

SAE A-21 Project Working Team (PWT) on Atmospheric Absorption,

As most of you are aware the U.S. Department of Transportation, John A. Volpe National Transportation Systems Center, Acoustics Facility (Volpe Center), in support of the Federal Aviation Administration's (FAA) Office of Environment and Energy (AEE), have been working under the auspices of a Society of Automotive Engineers' (SAE) Project Working Team (PWT) to assist in the development of an update to SAE Aerospace Recommended Practice (ARP) 866A, *Standard Values of Atmospheric Absorption as a Function of Temperature and Humidity* [REF 1]. Other members of the PWT include Airbus Industries, Boeing Commercial Airplane Company, Dytec Engineering, Mestre Grieve and Associates, and Louis Sutherland.

The work focus for the PWT has been an evaluation of the equations included in two standards, ISO 9613-1, *Acoustics - Attenuation of sound during propagation outdoors - Part 1: Calculation of the absorption of sound by the atmosphere* [REF 2], and ANSI S1.26-1995, *Standard Values of Atmospheric Absorption as a Function of Temperature and Humidity* [REF 3]. The ISO and ANSI equations for computing sound attenuation rates as a function of propagation distance are algebraically identical to one another, and specify computation as a function of temperature, relative humidity, and atmospheric pressure for a **single, discrete frequency or pure-tone.**

To date, the PWT has sufficient data and has performed the necessary analyses to conclude that: (1) the pure-tone absorption equations developed in support of the ANSI and ISO standards are more accurate than the equations of SAE ARP 866A, and (2) adoption of the new equations would result in a change in noise certificated levels for aircraft certified under current FAA and International Civil Aviation Organization (ICAO) noise regulations. These conclusions are based on the results presented in References 4 and 5.

However, the noise certification regulations of both the FAA and ICAO, which mandate the use of SAE ARP 866A, require that noise data be analyzed in one-third-octave bands. Application of the pure-tone sound absorption equations to one-third octave-band data is the only remaining technical obstacle to the adoption of the ISO and ANSI pure-tone equations as a replacement to SAE ARP 866A.

The authors of the ISO and ANSI standards included methods for adapting the pure-tone algorithms for use in fractional-octave-band analysis, e.g., a one-third octave-band or full octave-band analysis. Annex D of both the ISO and ANSI standard present a relatively complex but technically sound method for adapting the pure-tone algorithms for use in fractional octave-band analysis -- the spectrum integrated or **exact method**. In addition, Annex E of the ANSI standard presents a more empirical method of adapting the pure-tone algorithms to one-third-octave bands -- the **approximate method**. For the purpose of aircraft noise certification, the exact method is considered too complex. The approximate method, although easy to implement, is cited in Annex E as accurate only for total path-length absorptions of less than 50 dB. Such a limitation is not adequate for aircraft noise certification, where the total path-length absorption can exceed several hundred decibels, depending on temperature, relative humidity, and source-to-receptor distance. It is particularly not appropriate fro the development of noise-power-distance (NPD) data.

This memorandum presents a brief overview of a newly-proposed approach for adapting the pure-tone algorithms in the ISO and ANSI standards to one-third octave-band data - the **proposed method**. The proposed method can be considered an extension to the approximate method included in Annex E of the ANSI standard. In that regard, the proposed method produces results, which are almost identical to the ANSI approximate method up to 50 dB path-length absorption (see error comparison below). However, where the approximate method breaks down after about 50 dB path-length absorption, the proposed method is considered reliable to 200 dB total path-length absorption. Consequently, the proposed method is considered appropriate for both the adjustment of aircraft noise certification data and for development of NPD data.

The proposed method was developed using an approach similar to that used in the error assessment of the ANSI approximate method [REF 6]. Specifically, the exact method presented in both the ISO and ANSI standards was considered the reference or "gold standard" by which the proposed method was judged. This is considered a reasonable approach since the exact method effectively derives pure-tone sound pressure levels (SPL) from one-third octave-band data. The pure-tone absorption equations are then applied to the derived SPL data. An inherent assumption in this process is that of the one-third octave-band filter shape. For the development of the proposed method a third-order Butterworth filter shape was assumed. This was the same assumption made by the authors of the ANSI approximate method, and it appears to be reasonable since most analyzer manufacturers base their filter design on the traditional third-order Butterworth.

In developing the proposed method a set of broadband (non-tonal) spectra were processed over the complete range of temperature and relative humidity conditions allowed under aircraft noise certification. Note that although the pure-tone ISO and ANSI equations include atmospheric pressure as a variable, previous studies have shown data to be negligibly effected by pressure [REF 5 & 6]. Hence for the purpose of the development of the proposed method, atmospheric pressure was set to a fixed, ISA, sea level value of 1013.25 mm Hg. Data were processed in two ways: (1) from the source to a fixed receiver distance (Case 1); and (2) from a fixed receiver distance to the source (Case 2). In each case the result was a set of points and associated curves relating attenuation by absorption, computed at mid-band for each one-third octave-band frequency and the exact band-level attenuation computed for that band. Figures 1 and 2 respectively show a set of curves for Case 1 and Case 2 computed at 25°C and 70% relative humidity.

Using the computer program STATISTICA, a statistical analysis was then performed with a goal of developing an empirical equation to compute representative attenuation by absorption for a particular one-third-octave band. The general form of the equation for the ANSI approximate method was used as a starting point. The goal of the analysis was to extend the applicability of that equation to well beyond the recommended 50 dB path-length absorption, using the data derived from the exact method.

Initially two statistical equations were developed, one for Case 1 and a second for Case 2. Although the case dependency on the equations is apparent, the need for reciprocity dictated that a single equation be maintained. That is to say, if two separate equations were maintained data could not be corrected from a receiver to the source and back to the receiver, under the same temperature and humidity conditions such that the original starting data (at a receiver) could be replicated. Consequently, it was decided to maintain the Case 1 equation since it would result in smaller errors as compared with the Case 2 equation. The Case 1 equation was also more consistent with the ANSI approximate method up to 50 dB path-length absorption. As noted in Figures 1 and 2, the band-level attenuation versus the mid-band-level attenuation is data slope dependant; thus, the equations were developed with a slope term. The introduction of a slope term in the equation would however also preclude reciprocity since, for example, the data would be corrected from a receiver to the source assuming one slope and then corrected back to the receiver assuming a different slope. Consequently, the final version of the proposed equation assumes a fixed slope of 3 dB per one-third-octave band. This was found to be a good compromise with the test data and considered to be fairly representative for most broadband aircraft spectra.

The equation for the proposed method for predicting attenuation by atmospheric absorption is as follows:

**Predicted Attenuation = (A + B*Mid)*(1+C*(D - E*Mid))^F**     (1)

where Mid = Band-Level Attenuation at the exact mid-band frequency and A= -0.02397; B= 0.867941757; C= 0.111761; D= 0.95824; E= 0.008191; F= 1.6.

Based on a comparison with data corrected using the exact method, the errors associated with the proposed method and the errors associated with the ANSI approximate method are as follows:

| Attenuation Range (dB) | Error Proposed Method (dB) | Error ANSI Approximate Method (dB) |
|---|---|---|
| 0 – 10 | 0.5, 0 | 0.3, -0.3 |
| 10 – 50 | 2.5, -0.1 | 0.6, -5 |
| 50 – 100 | 2.5, -4 | -5, -10 |
| 100 – 150 | 2.0, -11 | -10, -25 |

                    150 – 200                     -2.0, -20                     -25,-50

As you can see there is a substantial improvement in accuracy and in the useful range to be realized with the proposed method. I have included with this memorandum an MS Excel spreadsheet, which can be used to assess the proposed method using example spectral data. The spreadsheet is self-explanatory and is setup to mimic a simplified procedure under current FAA and ICAO noise regulations. The Volpe Center is currently using this spreadsheet to exhaustively test the proposed method. I am requesting the addressees of this memorandum to also exercise the method with your own data and provide feedback to me by March 29, 2002.

Sincerely Yours,

Gregg G. Fleming

cc: J. Brooks, SAE, A-21 Chair
    M. Marsan, FAA/AEE
    J. Gulding, FAA/AEE

## B.1.1 References

[1] Society of Automotive Engineers, Committee A-21, Aircraft Noise, <u>Standard Values of Atmospheric Absorption as a Function of Temperature and Humidity</u>, Aerospace Recommended Practice No. 866A, Warrendale, PA: Society of Automotive Engineers, Inc., March 1975.

[2] International Organization for Standardization, Committee ISO/TC 43, Acoustics, Sub-Committee SC 1, Noise, <u>Acoustics - Attenuation of sound during propagation outdoors - Part 1: Calculation of the absorption of sound by the atmosphere</u>, ISO 9613-1, Geneva, Switzerland: International Organization for Standardization, 1993.

[3] American National Standards Institute, Committee S1, Acoustics, <u>Method for Calculation of the Absorption of Sound by the Atmosphere</u>, ANSI S1.26-1995, New York, NY: American National Standards Institute, September 1995.

[4] Rickley, E.J., and Fleming, G.G., <u>Computing the Absorption of Sound by the Atmosphere and its Applicability to Aircraft Noise Certification</u>, Cambridge, MA: Volpe Center, August 1998.

[5] Lempereur, Pierre, Presentation to SAE A-21: <u>Modeling of Sound Attenuation By Atmosphere, Application to Aircraft Noise Certification</u>, Toulouse, France: Aerospatiale, October 1998.

[6] Joppa, P.D., et.al., *Representative frequency approach to the effect of bandpass*

*filters on evaluation of sound attenuation by the atmosphere*, Noise Control Engineering Journal 44(6) 1996 Nov-Dec.

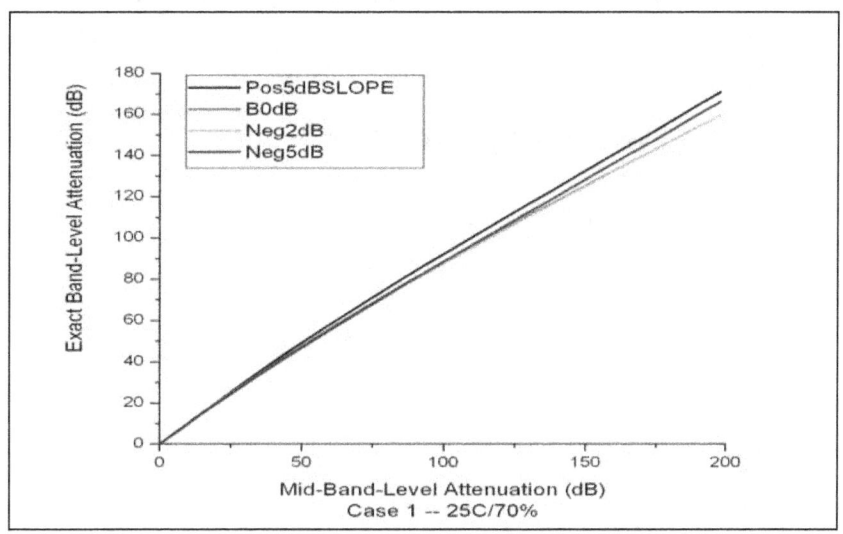

**Figure B.1** – Case 1 -- 25°C/70%RH Exact Band-Level vs. Mid-Band-Level Attenuation Data Slopes +5, 0, -2, and -5 dB

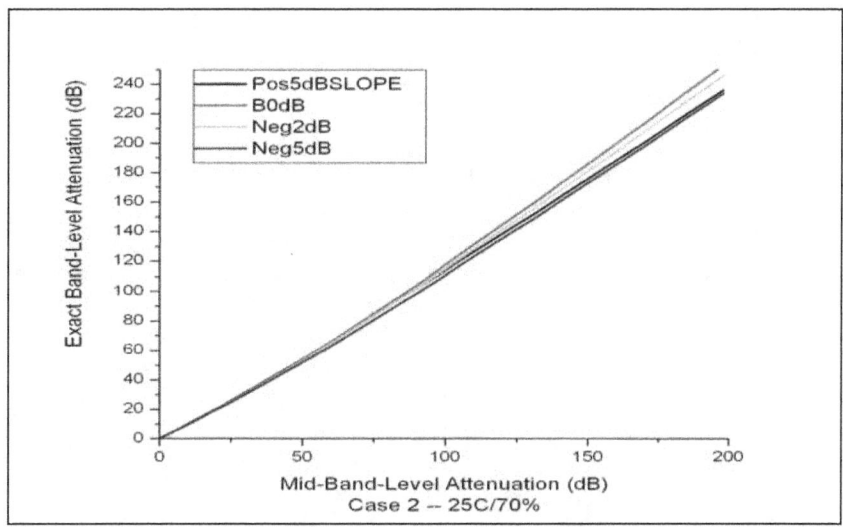

**Figure B.2** – Case 2 -- 25°C/70%RH Exact Band-Level vs. Mid-Band-Level Attenuation Data Slopes +5, 0, -2, and -5 dB

**B.2 Addendum, March 15, 2002**

The refined equation in the second version of the proposed method for predicting attenuation by atmospheric absorption is the same as in the memorandum above in cases where Band-Level Attenuation at the exact mid-band frequency, or Mid, is less than 150

**Predicted Attenuation = (A + B\*Mid)\*(1+C\*(D - E\*Mid))^F** (1)

where A, B, C, D, E, and F retain the same values as above, A= -0.02397; B= 0.867941757; C= 0.111761; D= 0.95824; E= 0.008191, and F= 1.6.

However, in cases where Mid ≥ 150, the equation changes to

**Predicted Attenuation = ((A+B\*150)\*(1+C\*(D-E\*150))^F)\*0.95\*Mid/150**   (2)

where A, B, C, D, E, and F retain the same values.

## Appendix C: $L_{ASmx}$ Difference Data Tables

Appendix C presents 105 tables containing the $L_{ASmx}$ vs. distance data generated with 866A and Version 2 of the ISO/ANSI spreadsheet, using the 21 as-measured spectra and the five different temperature/RH combinations.

### C.1 As-Measured Spectra Adjusted to 77°F/70% RH

**Table C.1** Spectral Class 101 at 77°F/70% RH.

| Distance (ft) | 866A (dBA) | ISO/ANSI (dBA) | Difference (ISO/ANSI - 866A) (dBA) |
|---|---|---|---|
| 200 | 94.9 | 94.9 | 0.0 |
| 400 | 88.5 | 88.5 | 0.0 |
| 630 | 84.1 | 84.1 | 0.0 |
| 1000 | 79.6 | 79.5 | 0.0 |
| 2000 | 72.2 | 72.2 | 0.0 |
| 4000 | 64.0 | 64.0 | 0.0 |
| 6300 | 58.1 | 58.2 | 0.1 |
| 10000 | 51.6 | 52.1 | 0.5 |
| 16000 | 44.3 | 45.7 | 1.3 |
| 25000 | 36.8 | 39.4 | 2.7 |

**Table C.2** Spectral Class 102 at 77°F/70% RH.

| Distance (ft) | 866A (dBA) | ISO/ANSI (dBA) | Difference (ISO/ANSI - 866A) (dBA) |
|---|---|---|---|
| 200 | 94.7 | 94.8 | 0.1 |
| 400 | 88.2 | 88.2 | 0.0 |
| 630 | 83.8 | 83.81 | 0.0 |
| 1000 | 79.1 | 79.1 | 0.0 |
| 2000 | 71.5 | 71.5 | -0.1 |
| 4000 | 62.9 | 62.7 | -0.2 |
| 6300 | 56.4 | 56.2 | -0.2 |
| 10000 | 49.0 | 48.9 | -0.1 |
| 16000 | 40.5 | 41.2 | 0.7 |
| 25000 | 31.9 | 34.0 | 2.2 |

**Table C.3** Spectral Class 103 at 77°F/70% RH.

| Distance (ft) | 866A (dBA) | ISO/ANSI (dBA) | Difference (ISO/ANSI - 866A) (dBA) |
|---|---|---|---|
| 200 | 96.7 | 96.5 | -0.2 |
| 400 | 90.1 | 89.9 | -0.1 |
| 630 | 85.5 | 85.4 | -0.1 |
| 1000 | 80.6 | 80.6 | 0.0 |
| 2000 | 72.6 | 72.7 | 0.1 |
| 4000 | 63.7 | 63.8 | 0.1 |
| 6300 | 57.5 | 57.7 | 0.1 |
| 10000 | 50.8 | 51.3 | 0.5 |
| 16000 | 43.3 | 44.6 | 1.2 |
| 25000 | 35.4 | 37.9 | 2.5 |

**Table C.4** Spectral Class 104 at 77°F/70% RH.

| Distance (ft) | 866A (dBA) | ISO/ANSI (dBA) | Difference (ISO/ANSI - 866A) (dBA) |
|---|---|---|---|
| 200 | 93.9 | 93.9 | 0.0 |
| 400 | 87.5 | 87.5 | 0.0 |
| 630 | 83.2 | 83.1 | 0.0 |
| 1000 | 78.5 | 78.5 | 0.0 |
| 2000 | 71.0 | 71.0 | 0.0 |
| 4000 | 62.5 | 62.4 | -0.1 |
| 6300 | 56.2 | 56.2 | 0.0 |
| 10000 | 49.2 | 49.4 | 0.2 |
| 16000 | 41.2 | 42.0 | 0.9 |
| 25000 | 32.5 | 34.7 | 2.2 |

**Table C.5** Spectral Class 105 at 77°F/70% RH.

| Distance (ft) | 866A (dBA) | ISO/ANSI (dBA) | Difference (ISO/ANSI - 866A) (dBA) |
|---|---|---|---|
| 200 | 95.8 | 95.7 | -0.1 |
| 400 | 89.3 | 89.2 | -0.1 |
| 630 | 84.8 | 84.8 | -0.1 |
| 1000 | 80.1 | 80.1 | 0.0 |
| 2000 | 72.5 | 72.5 | 0.0 |
| 4000 | 64.1 | 64.1 | 0.0 |
| 6300 | 58.1 | 58.2 | 0.1 |
| 10000 | 51.5 | 52.0 | 0.5 |
| 16000 | 44.2 | 45.5 | 1.3 |
| 25000 | 36.5 | 39.2 | 2.6 |

**Table C.6** Spectral Class 106 at 77°F/70% RH.

| Distance (ft) | 866A (dBA) | ISO/ANSI (dBA) | Difference (ISO/ANSI - 866A) (dBA) |
|---|---|---|---|
| 200 | 96.2 | 96.2 | 0.0 |
| 400 | 89.7 | 89.6 | 0.0 |
| 630 | 85.2 | 85.2 | 0.0 |
| 1000 | 80.5 | 80.5 | 0.0 |
| 2000 | 72.9 | 72.9 | 0.0 |
| 4000 | 64.5 | 64.4 | -0.1 |
| 6300 | 58.3 | 58.2 | -0.1 |
| 10000 | 51.2 | 51.3 | 0.1 |
| 16000 | 43.1 | 43.8 | 0.8 |
| 25000 | 34.4 | 36.6 | 2.2 |

**Table C.7** Spectral Class 107 at 77°F/70% RH.

| Distance (ft) | 866A (dBA) | ISO/ANSI (dBA) | Difference (ISO/ANSI - 866A) (dBA) |
|---|---|---|---|
| 200 | 97.9 | 97.6 | -0.3 |
| 400 | 91.2 | 91.0 | -0.2 |
| 630 | 86.5 | 86.4 | -0.1 |
| 1000 | 81.4 | 81.4 | 0.0 |
| 2000 | 73.1 | 73.2 | 0.1 |
| 4000 | 63.9 | 64.0 | 0.1 |
| 6300 | 57.5 | 57.5 | 0.0 |
| 10000 | 50.6 | 50.9 | 0.3 |
| 16000 | 42.9 | 43.9 | 1.1 |
| 25000 | 34.8 | 37.2 | 2.5 |

**Table C.8** Spectral Class 108 at 77°F/70% RH.

| Distance (ft) | 866A (dBA) | ISO/ANSI (dBA) | Difference (ISO/ANSI - 866A) (dBA) |
|---|---|---|---|
| 200 | 95.4 | 95.7 | 0.2 |
| 400 | 88.6 | 88.6 | 0.0 |
| 630 | 83.9 | 83.8 | -0.1 |
| 1000 | 78.8 | 78.8 | 0.0 |
| 2000 | 70.7 | 70.7 | 0.0 |
| 4000 | 61.9 | 61.9 | 0.0 |
| 6300 | 55.6 | 55.6 | 0.0 |
| 10000 | 48.7 | 49.0 | 0.4 |
| 16000 | 41.3 | 42.5 | 1.3 |
| 25000 | 34.0 | 36.6 | 2.7 |

**Table C.9** Spectral Class 109 at 77°F/70% RH.

| Distance (ft) | 866A (dBA) | ISO/ANSI (dBA) | Difference (ISO/ANSI - 866A) (dBA) |
|---|---|---|---|
| 200 | 100.1 | 100.1 | 0.0 |
| 400 | 93.9 | 93.9 | 0.0 |
| 630 | 89.7 | 89.7 | 0.0 |
| 1000 | 85.4 | 85.4 | 0.0 |
| 2000 | 78.5 | 78.5 | 0.0 |
| 4000 | 71.0 | 71.0 | 0.0 |
| 6300 | 65.5 | 65.7 | 0.2 |
| 10000 | 59.4 | 60.0 | 0.6 |
| 16000 | 52.4 | 53.9 | 1.5 |
| 25000 | 45.1 | 47.9 | 2.8 |

**Table C.10** Spectral Class 110 at 77°F/70% RH.

| Distance (ft) | 866A (dBA) | ISO/ANSI (dBA) | Difference (ISO/ANSI - 866A) (dBA) |
|---|---|---|---|
| 200 | 100.3 | 100.3 | 0.0 |
| 400 | 93.9 | 93.9 | 0.0 |
| 630 | 89.7 | 89.6 | 0.0 |
| 1000 | 85.2 | 85.2 | 0.0 |
| 2000 | 78.3 | 78.3 | 0.0 |
| 4000 | 70.9 | 71.0 | 0.2 |
| 6300 | 65.5 | 65.9 | 0.3 |
| 10000 | 59.5 | 60.2 | 0.7 |
| 16000 | 52.6 | 54.0 | 1.4 |
| 25000 | 44.9 | 47.4 | 2.6 |

**Table C.11** Spectral Class 111 at 77°F/70% RH.

| Distance (ft) | 866A (dBA) | ISO/ANSI (dBA) | Difference (ISO/ANSI - 866A) (dBA) |
|---|---|---|---|
| 200 | 106.2 | 106.1 | -0.1 |
| 400 | 100.1 | 100.0 | 0.0 |
| 630 | 96.0 | 95.9 | 0.0 |
| 1000 | 91.8 | 91.8 | 0.0 |
| 2000 | 85.3 | 85.4 | 0.1 |
| 4000 | 78.4 | 78.6 | 0.3 |
| 6300 | 73.4 | 73.9 | 0.5 |
| 10000 | 68.0 | 68.9 | 0.9 |
| 16000 | 61.7 | 63.3 | 1.6 |
| 25000 | 55.0 | 57.7 | 2.7 |

**Table C.12** Spectral Class 112 at 77°F/70% RH.

| Distance (ft) | 866A (dBA) | ISO/ANSI (dBA) | Difference (ISO/ANSI - 866A) (dBA) |
|---|---|---|---|
| 200 | 100.2 | 100.1 | -0.1 |
| 400 | 93.8 | 93.7 | -0.1 |
| 630 | 89.5 | 89.4 | -0.1 |
| 1000 | 85.0 | 85.0 | 0.0 |
| 2000 | 77.9 | 78.0 | 0.1 |
| 4000 | 70.7 | 71.0 | 0.3 |
| 6300 | 65.8 | 66.3 | 0.5 |
| 10000 | 60.6 | 61.6 | 1.0 |
| 16000 | 55.0 | 56.8 | 1.8 |
| 25000 | 49.0 | 51.9 | 2.9 |

**Table C.13** Spectral Class 113 at 77°F/70% RH.

| Distance (ft) | 866A (dBA) | ISO/ANSI (dBA) | Difference (ISO/ANSI - 866A) (dBA) |
|---|---|---|---|
| 200 | 94.6 | 94.5 | -0.1 |
| 400 | 88.2 | 88.1 | 0.0 |
| 630 | 83.8 | 83.7 | 0.0 |
| 1000 | 79.1 | 79.0 | 0.0 |
| 2000 | 71.5 | 71.4 | 0.0 |
| 4000 | 62.9 | 62.8 | -0.1 |
| 6300 | 56.6 | 56.5 | -0.1 |
| 10000 | 49.4 | 49.5 | 0.1 |
| 16000 | 41.2 | 42.1 | 0.8 |
| 25000 | 32.5 | 34.7 | 2.2 |

**Table C.14** Spectral Class 116 at 77°F/70% RH.

| Distance (ft) | 866A (dBA) | ISO/ANSI (dBA) | Difference (ISO/ANSI - 866A) (dBA) |
|---|---|---|---|
| 200 | 95.6 | 95.6 | 0.0 |
| 400 | 89.3 | 89.3 | 0.0 |
| 630 | 85.1 | 85.0 | 0.0 |
| 1000 | 80.6 | 80.5 | 0.0 |
| 2000 | 73.3 | 73.3 | 0.0 |
| 4000 | 65.3 | 65.2 | -0.1 |
| 6300 | 59.3 | 59.3 | 0.0 |
| 10000 | 52.5 | 52.7 | 0.2 |
| 16000 | 44.6 | 45.6 | 1.0 |
| 25000 | 36.2 | 38.6 | 2.4 |

**Table C.15** Spectral Class 203 at 77°F/70% RH.

| Distance (ft) | 866A (dBA) | ISO/ANSI (dBA) | Difference (ISO/ANSI - 866A) (dBA) |
|---|---|---|---|
| 200 | 98.7 | 98.4 | -0.3 |
| 400 | 91.8 | 91.5 | -0.2 |
| 630 | 86.9 | 86.7 | -0.2 |
| 1000 | 81.5 | 81.5 | 0.0 |
| 2000 | 72.8 | 72.9 | 0.1 |
| 4000 | 63.6 | 63.6 | 0.0 |
| 6300 | 57.2 | 57.1 | -0.1 |
| 10000 | 50.0 | 50.2 | 0.2 |
| 16000 | 42.0 | 43.0 | 1.0 |
| 25000 | 33.6 | 36.0 | 2.4 |

**Table C.16** Spectral Class 205 at 77°F/70% RH.

| Distance (ft) | 866A (dBA) | ISO/ANSI (dBA) | Difference (ISO/ANSI - 866A) (dBA) |
|---|---|---|---|
| 200 | 95.3 | 95.2 | 0.0 |
| 400 | 88.7 | 88.6 | -0.1 |
| 630 | 84.1 | 84.0 | -0.1 |
| 1000 | 79.2 | 79.2 | 0.0 |
| 2000 | 71.4 | 71.4 | 0.0 |
| 4000 | 62.7 | 62.7 | -0.1 |
| 6300 | 56.5 | 56.4 | 0.0 |
| 10000 | 49.5 | 49.8 | 0.3 |
| 16000 | 41.9 | 43.0 | 1.2 |
| 25000 | 34.0 | 36.6 | 2.6 |

**Table C.17** Spectral Class 219 at 77°F/70% RH.

| Distance (ft) | 866A (dBA) | ISO/ANSI (dBA) | Difference (ISO/ANSI - 866A) (dBA) |
|---|---|---|---|
| 200 | 97.6 | 97.6 | 0.0 |
| 400 | 91.4 | 91.3 | 0.0 |
| 630 | 87.2 | 87.1 | 0.0 |
| 1000 | 82.8 | 82.8 | 0.0 |
| 2000 | 75.9 | 75.9 | 0.0 |
| 4000 | 68.4 | 68.5 | 0.1 |
| 6300 | 63.0 | 63.2 | 0.3 |
| 10000 | 56.9 | 57.6 | 0.7 |
| 16000 | 50.1 | 51.5 | 1.5 |
| 25000 | 42.6 | 45.4 | 2.8 |

**Table C.18** Spectral Class 222 at 77°F/70% RH.

| Distance (ft) | 866A (dBA) | ISO/ANSI (dBA) | Difference (ISO/ANSI - 866A) (dBA) |
|---|---|---|---|
| 200 | 103.9 | 103.9 | 0.0 |
| 400 | 97.8 | 97.7 | 0.0 |
| 630 | 93.6 | 93.6 | 0.0 |
| 1000 | 89.3 | 89.3 | 0.0 |
| 2000 | 82.6 | 82.6 | 0.0 |
| 4000 | 75.2 | 75.3 | 0.1 |
| 6300 | 69.8 | 70.0 | 0.2 |
| 10000 | 63.6 | 64.1 | 0.5 |
| 16000 | 56.2 | 57.4 | 1.2 |
| 25000 | 48.0 | 50.3 | 2.3 |

**Table C.19** Spectral Class 302 at 77°F/70% RH.

| Distance (ft) | 866A (dBA) | ISO/ANSI (dBA) | Difference (ISO/ANSI - 866A) (dBA) |
|---|---|---|---|
| 200 | 93.6 | 93.5 | -0.1 |
| 400 | 87.1 | 86.9 | -0.1 |
| 630 | 82.5 | 82.4 | -0.1 |
| 1000 | 77.6 | 77.6 | 0.0 |
| 2000 | 69.4 | 69.4 | 0.0 |
| 4000 | 60.0 | 60.0 | 0.0 |
| 6300 | 53.0 | 52.9 | -0.1 |
| 10000 | 45.3 | 45.5 | 0.2 |
| 16000 | 37.1 | 38.1 | 1.0 |
| 25000 | 28.9 | 31.3 | 2.4 |

**Table C.20** Spectral Class 307 at 77°F/70% RH.

| Distance (ft) | 866A (dBA) | ISO/ANSI (dBA) | Difference (ISO/ANSI - 866A) (dBA) |
|---|---|---|---|
| 200 | 97.4 | 97.3 | -0.1 |
| 400 | 91.0 | 91.0 | 0.0 |
| 630 | 86.8 | 86.7 | 0.0 |
| 1000 | 82.3 | 82.2 | 0.0 |
| 2000 | 75.2 | 75.2 | 0.0 |
| 4000 | 67.5 | 67.6 | 0.1 |
| 6300 | 62.0 | 62.3 | 0.3 |
| 10000 | 56.0 | 56.7 | 0.7 |
| 16000 | 49.1 | 50.6 | 1.5 |
| 25000 | 41.7 | 44.5 | 2.9 |

**Table C.21** MD900 Spectral Data at 77°F/70% RH.

| Distance (ft) | 866A (dBA) | ISO/ANSI (dBA) | Difference (ISO/ANSI - 866A) (dBA) |
|---|---|---|---|
| 200 | 81.3 | 81.2 | -0.1 |
| 400 | 74.9 | 74.8 | 0.0 |
| 630 | 70.5 | 70.5 | 0.0 |
| 1000 | 65.9 | 65.9 | 0.0 |
| 2000 | 58.5 | 58.5 | 0.0 |
| 4000 | 50.3 | 50.3 | 0.0 |
| 6300 | 44.4 | 44.5 | 0.1 |
| 10000 | 37.7 | 38.1 | 0.4 |
| 16000 | 30.1 | 31.2 | 1.1 |
| 25000 | 21.9 | 24.2 | 2.3 |

## C.2 As-Measured Spectra Adjusted to 85°F/35% RH

**Table C.22** Spectral Class 101 at 85°F/35% RH.

| Distance (ft) | 866A (dBA) | ISO/ANSI (dBA) | Difference (ISO/ANSI - 866A) (dBA) |
|---|---|---|---|
| 200 | 94.8 | 94.8 | 0.0 |
| 400 | 88.4 | 88.4 | 0.0 |
| 630 | 84.0 | 84.0 | 0.0 |
| 1000 | 79.4 | 79.4 | 0.0 |
| 2000 | 72.0 | 71.9 | 0.0 |
| 4000 | 63.7 | 63.5 | -0.1 |
| 6300 | 57.7 | 57.4 | -0.2 |
| 10000 | 51.0 | 50.8 | -0.2 |
| 16000 | 43.7 | 43.8 | 0.2 |
| 25000 | 36.0 | 37.0 | 1.0 |

**Table C.23** Spectral Class 102 at 85°F/35% RH.

| Distance (ft) | 866A (dBA) | ISO/ANSI (dBA) | Difference (ISO/ANSI - 866A) (dBA) |
|---|---|---|---|
| 200 | 94.7 | 94.6 | 0.0 |
| 400 | 88.1 | 88.1 | 0.0 |
| 630 | 83.7 | 83.6 | 0.0 |
| 1000 | 78.9 | 78.9 | 0.0 |
| 2000 | 71.3 | 71.2 | 0.0 |
| 4000 | 62.5 | 62.3 | -0.2 |
| 6300 | 55.9 | 55.5 | -0.4 |
| 10000 | 48.3 | 47.7 | -0.6 |
| 16000 | 39.7 | 39.3 | -0.4 |
| 25000 | 30.9 | 31.5 | 0.7 |

**Table C.24** Spectral Class 103 at 85°F/35% RH.

| Distance (ft) | 866A (dBA) | ISO/ANSI (dBA) | Difference (ISO/ANSI - 866A) (dBA) |
|---|---|---|---|
| 200 | 96.6 | 96.4 | -0.2 |
| 400 | 90.0 | 89.8 | -0.2 |
| 630 | 85.3 | 85.2 | -0.1 |
| 1000 | 80.4 | 80.3 | 0.0 |
| 2000 | 72.3 | 72.3 | 0.1 |
| 4000 | 63.3 | 63.3 | 0.0 |
| 6300 | 57.1 | 56.9 | -0.2 |
| 10000 | 50.2 | 49.9 | -0.3 |
| 16000 | 42.6 | 42.5 | -0.1 |
| 25000 | 34.5 | 35.0 | 0.6 |

**Table C.25** Spectral Class 104 at 85°F/35% RH.

| Distance (ft) | 866A (dBA) | ISO/ANSI (dBA) | Difference (ISO/ANSI - 866A) (dBA) |
|---|---|---|---|
| 200 | 93.9 | 93.9 | 0.0 |
| 400 | 87.5 | 87.4 | 0.0 |
| 630 | 83.1 | 83.1 | 0.0 |
| 1000 | 78.4 | 78.4 | 0.0 |
| 2000 | 70.8 | 70.8 | 0.0 |
| 4000 | 62.1 | 62.0 | -0.1 |
| 6300 | 55.7 | 55.4 | -0.3 |
| 10000 | 48.5 | 48.0 | -0.5 |
| 16000 | 40.3 | 39.8 | -0.5 |
| 25000 | 31.5 | 31.7 | 0.2 |

**Table C.26** Spectral Class 105 at 85°F/35% RH.

| Distance (ft) | 866A (dBA) | ISO/ANSI (dBA) | Difference (ISO/ANSI - 866A) (dBA) |
|---|---|---|---|
| 200 | 95.7 | 95.6 | -0.1 |
| 400 | 89.2 | 89.1 | -0.1 |
| 630 | 84.7 | 84.6 | -0.1 |
| 1000 | 79.9 | 79.9 | 0.0 |
| 2000 | 72.2 | 72.2 | 0.0 |
| 4000 | 63.7 | 63.6 | -0.1 |
| 6300 | 57.6 | 57.4 | -0.2 |
| 10000 | 51.0 | 50.7 | -0.2 |
| 16000 | 43.5 | 43.6 | 0.1 |
| 25000 | 35.7 | 36.6 | 1.0 |

**Table C.27** Spectral Class 106 at 85°F/35% RH.

| Distance (ft) | 866A (dBA) | ISO/ANSI (dBA) | Difference (ISO/ANSI - 866A) (dBA) |
|---|---|---|---|
| 200 | 96.1 | 96.1 | -0.1 |
| 400 | 89.6 | 89.5 | -0.1 |
| 630 | 85.1 | 85.0 | -0.1 |
| 1000 | 80.3 | 80.3 | 0.0 |
| 2000 | 72.6 | 72.6 | 0.0 |
| 4000 | 64.1 | 63.9 | -0.2 |
| 6300 | 57.8 | 57.3 | -0.4 |
| 10000 | 50.5 | 49.9 | -0.6 |
| 16000 | 42.2 | 41.7 | -0.5 |
| 25000 | 33.4 | 33.9 | 0.5 |

**Table C.28** Spectral Class 107 at 85°F/35% RH

| Distance (ft) | 866A (dBA) | ISO/ANSI (dBA) | Difference (ISO/ANSI - 866A) (dBA) |
|---|---|---|---|
| 200 | 97.8 | 97.5 | -0.3 |
| 400 | 91.1 | 90.8 | -0.3 |
| 630 | 86.3 | 86.1 | -0.2 |
| 1000 | 81.2 | 81.1 | -0.1 |
| 2000 | 72.7 | 72.7 | 0.0 |
| 4000 | 63.4 | 63.4 | -0.1 |
| 6300 | 57.0 | 56.7 | -0.3 |
| 10000 | 50.0 | 49.6 | -0.4 |
| 16000 | 42.1 | 41.9 | -0.2 |
| 25000 | 33.9 | 34.6 | 0.7 |

**Table C.29** Spectral Class 108 at 85°F/35% RH.

| Distance (ft) | 866A (dBA) | ISO/ANSI (dBA) | Difference (ISO/ANSI - 866A) (dBA) |
|---|---|---|---|
| 200 | 95.3 | 95.2 | 0.0 |
| 400 | 88.4 | 88.2 | -0.2 |
| 630 | 83.6 | 83.4 | -0.2 |
| 1000 | 78.5 | 78.4 | -0.1 |
| 2000 | 70.4 | 70.4 | 0.0 |
| 4000 | 61.5 | 61.5 | 0.0 |
| 6300 | 55.1 | 54.9 | -0.1 |
| 10000 | 48.1 | 48.0 | -0.1 |
| 16000 | 40.7 | 41.0 | 0.3 |
| 25000 | 33.2 | 34.6 | 1.4 |

**Table C.30** Spectral Class 109 at 85°F/35% RH.

| Distance (ft) | 866A (dBA) | ISO/ANSI (dBA) | Difference (ISO/ANSI - 866A) (dBA) |
|---|---|---|---|
| 200 | 100.1 | 100.1 | 0.0 |
| 400 | 93.8 | 93.8 | 0.0 |
| 630 | 89.7 | 89.6 | 0.0 |
| 1000 | 85.3 | 85.2 | -0.1 |
| 2000 | 78.4 | 78.3 | -0.1 |
| 4000 | 70.7 | 70.5 | -0.2 |
| 6300 | 65.2 | 64.9 | -0.2 |
| 10000 | 58.9 | 58.8 | -0.1 |
| 16000 | 51.8 | 52.1 | 0.3 |
| 25000 | 44.4 | 45.5 | 1.1 |

**Table C.31** Spectral Class 110 at 85°F/35% RH.

| Distance (ft) | 866A (dBA) | ISO/ANSI (dBA) | Difference (ISO/ANSI - 866A) (dBA) |
|---|---|---|---|
| 200 | 100.2 | 100.2 | 0.0 |
| 400 | 93.9 | 93.8 | -0.1 |
| 630 | 89.6 | 89.5 | -0.1 |
| 1000 | 85.1 | 85.0 | -0.1 |
| 2000 | 78.2 | 78.0 | -0.1 |
| 4000 | 70.6 | 70.4 | -0.2 |
| 6300 | 65.2 | 64.9 | -0.3 |
| 10000 | 59.1 | 58.8 | -0.3 |
| 16000 | 51.9 | 51.8 | -0.1 |
| 25000 | 44.0 | 44.5 | 0.5 |

**Table C.32** Spectral Class 111 at 85°F/35% RH.

| Distance (ft) | 866A (dBA) | ISO/ANSI (dBA) | Difference (ISO/ANSI - 866A) (dBA) |
|---|---|---|---|
| 200 | 106.2 | 106.1 | -0.1 |
| 400 | 100.0 | 99.9 | -0.1 |
| 630 | 96.0 | 95.9 | -0.1 |
| 1000 | 91.7 | 91.6 | -0.1 |
| 2000 | 85.2 | 85.1 | -0.1 |
| 4000 | 78.2 | 78.1 | -0.1 |
| 6300 | 73.2 | 73.2 | 0.0 |
| 10000 | 67.6 | 67.7 | 0.1 |
| 16000 | 61.2 | 61.8 | 0.5 |
| 25000 | 54.3 | 55.7 | 1.4 |

**Table C.33** Spectral Class 112 at 85°F/35% RH.

| Distance (ft) | 866A (dBA) | ISO/ANSI (dBA) | Difference (ISO/ANSI - 866A) (dBA) |
|---|---|---|---|
| 200 | 100.2 | 100.0 | -0.2 |
| 400 | 93.8 | 93.6 | -0.1 |
| 630 | 89.4 | 89.3 | -0.1 |
| 1000 | 84.9 | 84.8 | -0.1 |
| 2000 | 77.8 | 77.8 | 0.0 |
| 4000 | 70.5 | 70.6 | 0.1 |
| 6300 | 65.6 | 65.9 | 0.3 |
| 10000 | 60.4 | 60.9 | 0.6 |
| 16000 | 54.6 | 55.7 | 1.1 |
| 25000 | 48.5 | 50.4 | 1.9 |

**Table C.34** Spectral Class 113 at 85°F/35% RH.

| Distance (ft) | 866A (dBA) | ISO/ANSI (dBA) | Difference (ISO/ANSI - 866A) (dBA) |
|---|---|---|---|
| 200 | 94.6 | 94.5 | -0.1 |
| 400 | 88.1 | 88.0 | -0.1 |
| 630 | 83.6 | 83.6 | 0.0 |
| 1000 | 78.9 | 78.9 | 0.0 |
| 2000 | 71.2 | 71.2 | 0.0 |
| 4000 | 62.5 | 62.4 | -0.1 |
| 6300 | 56.1 | 55.7 | -0.4 |
| 10000 | 48.7 | 48.2 | -0.6 |
| 16000 | 40.4 | 39.9 | -0.5 |
| 25000 | 31.5 | 31.7 | 0.2 |

**Table C.35** Spectral Class 116 at 85°F/35% RH.

| Distance (ft) | 866A (dBA) | ISO/ANSI (dBA) | Difference (ISO/ANSI - 866A) (dBA) |
|---|---|---|---|
| 200 | 95.6 | 95.6 | 0.0 |
| 400 | 89.3 | 89.3 | 0.0 |
| 630 | 85.0 | 85.0 | 0.0 |
| 1000 | 80.4 | 80.4 | 0.0 |
| 2000 | 73.1 | 73.0 | -0.1 |
| 4000 | 64.9 | 64.7 | -0.2 |
| 6300 | 58.9 | 58.4 | -0.4 |
| 10000 | 51.9 | 51.3 | -0.5 |
| 16000 | 43.8 | 43.5 | -0.4 |
| 25000 | 35.2 | 35.6 | 0.4 |

**Table C.36** Spectral Class 203 at 85°F/35% RH.

| Distance (ft) | 866A (dBA) | ISO/ANSI (dBA) | Difference (ISO/ANSI - 866A) (dBA) |
|---|---|---|---|
| 200 | 98.6 | 98.1 | -0.4 |
| 400 | 91.5 | 91.1 | -0.4 |
| 630 | 86.5 | 86.2 | -0.3 |
| 1000 | 81.1 | 80.9 | -0.2 |
| 2000 | 72.4 | 72.3 | 0.0 |
| 4000 | 63.1 | 63.0 | -0.1 |
| 6300 | 56.6 | 56.3 | -0.3 |
| 10000 | 49.4 | 48.9 | -0.5 |
| 16000 | 41.2 | 40.9 | -0.3 |
| 25000 | 32.6 | 33.1 | 0.5 |

**Table C.37** Spectral Class 205 at 85°F/35% RH.

| Distance (ft) | 866A (dBA) | ISO/ANSI (dBA) | Difference (ISO/ANSI - 866A) (dBA) |
|---|---|---|---|
| 200 | 95.2 | 95.1 | -0.1 |
| 400 | 88.5 | 88.4 | -0.1 |
| 630 | 83.9 | 83.8 | -0.1 |
| 1000 | 79.0 | 79.0 | -0.1 |
| 2000 | 71.1 | 71.1 | 0.0 |
| 4000 | 62.3 | 62.3 | -0.1 |
| 6300 | 56.0 | 55.7 | -0.2 |
| 10000 | 48.9 | 48.6 | -0.3 |
| 16000 | 41.2 | 41.2 | 0.1 |
| 25000 | 33.2 | 34.2 | 1.0 |

**Table C.38** Spectral Class 219 at 85°F/35% RH.

| Distance (ft) | 866A (dBA) | ISO/ANSI (dBA) | Difference (ISO/ANSI - 866A) (dBA) |
|---|---|---|---|
| 200 | 97.6 | 97.5 | 0.0 |
| 400 | 91.3 | 91.3 | 0.0 |
| 630 | 87.1 | 87.1 | -0.1 |
| 1000 | 82.7 | 82.6 | -0.1 |
| 2000 | 75.7 | 75.6 | -0.1 |
| 4000 | 68.1 | 68.0 | -0.2 |
| 6300 | 62.6 | 62.4 | -0.2 |
| 10000 | 56.5 | 56.3 | -0.1 |
| 16000 | 49.4 | 49.7 | 0.2 |
| 25000 | 41.8 | 42.9 | 1.0 |

**Table C.39** Spectral Class 222 at 85°F/35% RH.

| Distance (ft) | 866A (dBA) | ISO/ANSI (dBA) | Difference (ISO/ANSI - 866A) (dBA) |
|---|---|---|---|
| 200 | 103.9 | 103.9 | 0.0 |
| 400 | 97.7 | 97.7 | -0.1 |
| 630 | 93.6 | 93.5 | -0.1 |
| 1000 | 89.3 | 89.2 | -0.1 |
| 2000 | 82.5 | 82.3 | -0.2 |
| 4000 | 75.0 | 74.7 | -0.3 |
| 6300 | 69.5 | 69.0 | -0.4 |
| 10000 | 63.1 | 62.6 | -0.5 |
| 16000 | 55.5 | 55.1 | -0.4 |
| 25000 | 46.9 | 47.0 | 0.1 |

**Table C.40** Spectral Class 302 at 85°F/35% RH.

| Distance (ft) | 866A (dBA) | ISO/ANSI (dBA) | Difference (ISO/ANSI - 866A) (dBA) |
|---|---|---|---|
| 200 | 93.6 | 93.4 | -0.1 |
| 400 | 87.0 | 86.8 | -0.1 |
| 630 | 82.4 | 82.3 | -0.1 |
| 1000 | 77.4 | 77.4 | 0.0 |
| 2000 | 69.1 | 69.2 | 0.1 |
| 4000 | 59.5 | 59.6 | 0.1 |
| 6300 | 52.4 | 52.3 | 0.0 |
| 10000 | 44.6 | 44.4 | -0.3 |
| 16000 | 36.3 | 36.2 | -0.2 |
| 25000 | 27.9 | 28.6 | 0.6 |

**Table C.41** Spectral Class 307 at 85°F/35% RH.

| Distance (ft) | 866A (dBA) | ISO/ANSI (dBA) | Difference (ISO/ANSI - 866A) (dBA) |
|---|---|---|---|
| 200 | 97.3 | 97.2 | -0.1 |
| 400 | 91.0 | 90.9 | -0.1 |
| 630 | 86.7 | 86.6 | -0.1 |
| 1000 | 82.2 | 82.1 | -0.1 |
| 2000 | 75.0 | 74.9 | -0.1 |
| 4000 | 67.3 | 67.1 | -0.1 |
| 6300 | 61.7 | 61.5 | -0.2 |
| 10000 | 55.5 | 55.4 | -0.1 |
| 16000 | 48.5 | 48.8 | 0.3 |
| 25000 | 40.9 | 42.0 | 1.1 |

**Table C.42** MD900 Spectral Data at 85°F/35% RH.

| Distance (ft) | 866A (dBA) | ISO/ANSI (dBA) | Difference (ISO/ANSI - 866A) (dBA) |
|---|---|---|---|
| 200 | 81.2 | 81.1 | -0.1 |
| 400 | 74.8 | 74.7 | -0.1 |
| 630 | 70.4 | 70.3 | -0.1 |
| 1000 | 65.7 | 65.7 | 0.0 |
| 2000 | 58.3 | 58.2 | -0.1 |
| 4000 | 50.0 | 49.8 | -0.2 |
| 6300 | 44.0 | 43.6 | -0.4 |
| 10000 | 37.2 | 36.7 | -0.5 |
| 16000 | 29.4 | 29.0 | -0.3 |
| 25000 | 20.9 | 21.3 | 0.4 |

## C.3 As-Measured Spectra Adjusted to 85°F/85% RH

**Table C.43** Spectral Class 101 at 85°F/85% RH.

| Distance (ft) | 866A (dBA) | ISO/ANSI (dBA) | Difference (ISO/ANSI - 866A) (dBA) |
|---|---|---|---|
| 200 | 94.8 | 94.8 | 0.0 |
| 400 | 88.4 | 88.4 | 0.0 |
| 630 | 84.1 | 84.0 | -0.1 |
| 1000 | 79.4 | 79.4 | -0.1 |
| 2000 | 72.0 | 72.0 | 0.0 |
| 4000 | 63.7 | 63.9 | 0.2 |
| 6300 | 57.7 | 58.3 | 0.7 |
| 10000 | 51.0 | 52.5 | 1.5 |
| 16000 | 43.7 | 46.5 | 2.8 |
| 25000 | 36.0 | 40.6 | 4.6 |

**Table C.44** Spectral Class 102 at 85°F/85% RH.

| Distance (ft) | 866A (dBA) | ISO/ANSI (dBA) | Difference (ISO/ANSI - 866A) (dBA) |
|---|---|---|---|
| 200 | 94.7 | 94.8 | 0.1 |
| 400 | 88.2 | 88.2 | 0.0 |
| 630 | 83.7 | 83.7 | 0.0 |
| 1000 | 79.0 | 78.9 | 0.0 |
| 2000 | 71.3 | 71.2 | -0.1 |
| 4000 | 62.5 | 62.5 | 0.0 |
| 6300 | 55.9 | 56.2 | 0.3 |
| 10000 | 48.3 | 49.3 | 1.0 |
| 16000 | 39.7 | 42.1 | 2.4 |
| 25000 | 30.9 | 35.3 | 4.4 |

**Table C.45** Spectral Class 103 at 85°F/85% RH.

| Distance (ft) | 866A (dBA) | ISO/ANSI (dBA) | Difference (ISO/ANSI - 866A) (dBA) |
|---|---|---|---|
| 200 | 96.6 | 96.4 | -0.2 |
| 400 | 90.0 | 89.8 | -0.2 |
| 630 | 85.4 | 85.2 | -0.2 |
| 1000 | 80.4 | 80.3 | -0.2 |
| 2000 | 72.3 | 72.2 | -0.1 |
| 4000 | 63.3 | 63.5 | 0.2 |
| 6300 | 57.1 | 57.7 | 0.7 |
| 10000 | 50.2 | 51.8 | 1.6 |
| 16000 | 42.6 | 45.6 | 2.9 |
| 25000 | 34.5 | 39.3 | 4.8 |

**Table C.46** Spectral Class 104 at 85°F/85% RH.

| Distance (ft) | 866A (dBA) | ISO/ANSI (dBA) | Difference (ISO/ANSI - 866A) (dBA) |
|---|---|---|---|
| 200 | 93.9 | 93.8 | -0.1 |
| 400 | 87.5 | 87.4 | -0.1 |
| 630 | 83.1 | 83.0 | -0.1 |
| 1000 | 78.4 | 78.3 | -0.1 |
| 2000 | 70.8 | 70.7 | -0.1 |
| 4000 | 62.1 | 62.3 | 0.1 |
| 6300 | 55.7 | 56.3 | 0.6 |
| 10000 | 48.5 | 49.9 | 1.4 |
| 16000 | 40.3 | 43.1 | 2.8 |
| 25000 | 31.5 | 36.2 | 4.7 |

**Table C.47** Spectral Class 105 at 85°F/85% RH.

| Distance (ft) | 866A (dBA) | ISO/ANSI (dBA) | Difference (ISO/ANSI - 866A) (dBA) |
|---|---|---|---|
| 200 | 95.7 | 95.6 | -0.1 |
| 400 | 89.2 | 89.1 | -0.1 |
| 630 | 84.7 | 84.6 | -0.1 |
| 1000 | 79.9 | 79.9 | -0.1 |
| 2000 | 72.2 | 72.2 | 0.0 |
| 4000 | 63.7 | 64.0 | 0.2 |
| 6300 | 57.6 | 58.3 | 0.7 |
| 10000 | 51.0 | 52.5 | 1.5 |
| 16000 | 43.5 | 46.4 | 2.9 |
| 25000 | 35.7 | 40.4 | 4.7 |

**Table C.48** Spectral Class 106 at 85°F/85% RH.

| Distance (ft) | 866A (dBA) | ISO/ANSI (dBA) | Difference (ISO/ANSI - 866A) (dBA) |
|---|---|---|---|
| 200 | 96.2 | 96.2 | 0.0 |
| 400 | 89.6 | 89.6 | 0.0 |
| 630 | 85.1 | 85.1 | 0.0 |
| 1000 | 80.3 | 80.3 | 0.0 |
| 2000 | 72.7 | 72.7 | 0.0 |
| 4000 | 64.1 | 64.3 | 0.2 |
| 6300 | 57.8 | 58.3 | 0.5 |
| 10000 | 50.5 | 51.8 | 1.3 |
| 16000 | 42.2 | 44.8 | 2.6 |
| 25000 | 33.4 | 38.0 | 4.5 |

**Table C.49** Spectral Class 107 at 85°F/85% RH.

| Distance (ft) | 866A (dBA) | ISO/ANSI (dBA) | Difference (ISO/ANSI - 866A) (dBA) |
|---|---|---|---|
| 200 | 97.8 | 97.5 | -0.3 |
| 400 | 91.1 | 90.8 | -0.3 |
| 630 | 86.4 | 86.1 | -0.2 |
| 1000 | 81.2 | 81.0 | -0.2 |
| 2000 | 72.7 | 72.7 | -0.1 |
| 4000 | 63.5 | 63.6 | 0.2 |
| 6300 | 57.0 | 57.6 | 0.6 |
| 10000 | 50.0 | 51.4 | 1.4 |
| 16000 | 42.1 | 44.9 | 2.8 |
| 25000 | 33.9 | 38.5 | 4.7 |

**Table C.50** Spectral Class 108 at 85°F/85% RH.

| Distance (ft) | 866A (dBA) | ISO/ANSI (dBA) | Difference (ISO/ANSI - 866A) (dBA) |
|---|---|---|---|
| 200 | 95.4 | 95.7 | 0.4 |
| 400 | 88.5 | 88.6 | 0.1 |
| 630 | 83.7 | 83.8 | 0.0 |
| 1000 | 78.6 | 78.6 | 0.0 |
| 2000 | 70.4 | 70.4 | -0.1 |
| 4000 | 61.5 | 61.5 | 0.0 |
| 6300 | 55.1 | 55.5 | 0.5 |
| 10000 | 48.1 | 49.4 | 1.3 |
| 16000 | 40.7 | 43.3 | 2.6 |
| 25000 | 33.2 | 37.6 | 4.3 |

**Table C.51** Spectral Class 109 at 85°F/85% RH.

| Distance (ft) | 866A (dBA) | ISO/ANSI (dBA) | Difference (ISO/ANSI - 866A) (dBA) |
|---|---|---|---|
| 200 | 100.1 | 100.1 | 0.0 |
| 400 | 93.8 | 93.9 | 0.0 |
| 630 | 89.7 | 89.7 | 0.0 |
| 1000 | 85.3 | 85.4 | 0.1 |
| 2000 | 78.4 | 78.5 | 0.2 |
| 4000 | 70.7 | 71.2 | 0.4 |
| 6300 | 65.2 | 66.0 | 0.8 |
| 10000 | 58.9 | 60.5 | 1.6 |
| 16000 | 51.8 | 54.7 | 2.9 |
| 25000 | 44.4 | 49.0 | 4.7 |

**Table C.52** Spectral Class 110 at 85°F/85% RH.

| Distance (ft) | 866A (dBA) | ISO/ANSI (dBA) | Difference (ISO/ANSI - 866A) (dBA) |
|---|---|---|---|
| 200 | 100.2 | 100.3 | 0.1 |
| 400 | 93.9 | 93.9 | 0.0 |
| 630 | 89.6 | 89.6 | 0.0 |
| 1000 | 85.1 | 85.2 | 0.1 |
| 2000 | 78.2 | 78.4 | 0.2 |
| 4000 | 70.6 | 71.2 | 0.6 |
| 6300 | 65.2 | 66.3 | 1.0 |
| 10000 | 59.1 | 60.9 | 1.8 |
| 16000 | 51.9 | 55.0 | 3.1 |
| 25000 | 44.0 | 48.8 | 4.9 |

**Table C.53** Spectral Class 111 at 85°F/85% RH.

| Distance (ft) | 866A (dBA) | ISO/ANSI (dBA) | Difference (ISO/ANSI - 866A) (dBA) |
|---|---|---|---|
| 200 | 106.2 | 106.1 | 0.0 |
| 400 | 100.0 | 100.0 | 0.0 |
| 630 | 96.0 | 96.0 | 0.0 |
| 1000 | 91.7 | 91.8 | 0.1 |
| 2000 | 85.2 | 85.5 | 0.3 |
| 4000 | 78.2 | 78.8 | 0.6 |
| 6300 | 73.2 | 74.3 | 1.1 |
| 10000 | 67.6 | 69.4 | 1.8 |
| 16000 | 61.2 | 64.1 | 2.9 |
| 25000 | 54.3 | 58.7 | 4.4 |

**Table C.54** Spectral Class 112 at 85°F/85% RH.

| Distance (ft) | 866A (dBA) | ISO/ANSI (dBA) | Difference (ISO/ANSI - 866A) (dBA) |
|---|---|---|---|
| 200 | 100.2 | 100.0 | -0.1 |
| 400 | 93.8 | 93.6 | -0.1 |
| 630 | 89.4 | 89.3 | -0.1 |
| 1000 | 84.9 | 84.8 | 0.0 |
| 2000 | 77.8 | 77.9 | 0.1 |
| 4000 | 70.5 | 71.0 | 0.5 |
| 6300 | 65.6 | 66.5 | 0.9 |
| 10000 | 60.4 | 61.9 | 1.6 |
| 16000 | 54.6 | 57.2 | 2.6 |
| 25000 | 48.5 | 52.5 | 4.1 |

**Table C.55** Spectral Class 113 at 85°F/85% RH.

| Distance (ft) | 866A (dBA) | ISO/ANSI (dBA) | Difference (ISO/ANSI - 866A) (dBA) |
|---|---|---|---|
| 200 | 94.6 | 94.5 | -0.1 |
| 400 | 88.1 | 88.0 | -0.1 |
| 630 | 83.7 | 83.6 | -0.1 |
| 1000 | 78.9 | 78.8 | -0.1 |
| 2000 | 71.2 | 71.2 | -0.1 |
| 4000 | 62.5 | 62.6 | 0.1 |
| 6300 | 56.1 | 56.5 | 0.5 |
| 10000 | 48.7 | 50.0 | 1.3 |
| 16000 | 40.4 | 43.1 | 2.7 |
| 25000 | 31.5 | 36.2 | 4.7 |

**Table C.56** Spectral Class 116 at 85°F/85% RH.

| Distance (ft) | 866A (dBA) | ISO/ANSI (dBA) | Difference (ISO/ANSI - 866A) (dBA) |
|---|---|---|---|
| 200 | 95.6 | 95.6 | 0.0 |
| 400 | 89.3 | 89.3 | 0.0 |
| 630 | 85.0 | 85.0 | 0.0 |
| 1000 | 80.4 | 80.4 | 0.0 |
| 2000 | 73.1 | 73.2 | 0.1 |
| 4000 | 64.9 | 65.2 | 0.3 |
| 6300 | 58.9 | 59.5 | 0.7 |
| 10000 | 51.9 | 53.3 | 1.4 |
| 16000 | 43.8 | 46.6 | 2.8 |
| 25000 | 35.2 | 40.0 | 4.8 |

**Table C.57** Spectral Class 203 at 85°F/85% RH.

| Distance (ft) | 866A (dBA) | ISO/ANSI (dBA) | Difference (ISO/ANSI - 866A) (dBA) |
|---|---|---|---|
| 200 | 98.7 | 98.4 | -0.3 |
| 400 | 91.7 | 91.4 | -0.2 |
| 630 | 86.7 | 86.5 | -0.1 |
| 1000 | 81.2 | 81.2 | -0.1 |
| 2000 | 72.4 | 72.5 | 0.0 |
| 4000 | 63.2 | 63.3 | 0.1 |
| 6300 | 56.6 | 57.1 | 0.5 |
| 10000 | 49.4 | 50.7 | 1.3 |
| 16000 | 41.2 | 43.9 | 2.7 |
| 25000 | 32.6 | 37.4 | 4.7 |

**Table C.58** Spectral Class 205 at 85°F/85% RH.

| Distance (ft) | 866A (dBA) | ISO/ANSI (dBA) | Difference (ISO/ANSI - 866A) (dBA) |
|---|---|---|---|
| 200 | 95.2 | 95.2 | 0.0 |
| 400 | 88.6 | 88.5 | 0.0 |
| 630 | 84.0 | 83.9 | -0.1 |
| 1000 | 79.1 | 79.0 | -0.1 |
| 2000 | 71.1 | 71.1 | -0.1 |
| 4000 | 62.3 | 62.4 | 0.1 |
| 6300 | 56.0 | 56.5 | 0.5 |
| 10000 | 48.9 | 50.2 | 1.3 |
| 16000 | 41.2 | 43.9 | 2.7 |
| 25000 | 33.2 | 37.8 | 4.6 |

**Table C.59** Spectral Class 219 at 85°F/85% RH.

| Distance (ft) | 866A (dBA) | ISO/ANSI (dBA) | Difference (ISO/ANSI - 866A) (dBA) |
|---|---|---|---|
| 200 | 97.6 | 97.5 | 0.0 |
| 400 | 91.3 | 91.3 | 0.0 |
| 630 | 87.1 | 87.1 | 0.0 |
| 1000 | 82.7 | 82.7 | 0.0 |
| 2000 | 75.7 | 75.9 | 0.2 |
| 4000 | 68.1 | 68.6 | 0.5 |
| 6300 | 62.6 | 63.6 | 0.9 |
| 10000 | 56.5 | 58.2 | 1.7 |
| 16000 | 49.4 | 52.4 | 3.0 |
| 25000 | 41.8 | 46.6 | 4.8 |

**Table C.60** Spectral Class 222 at 85°F/85% RH.

| Distance (ft) | 866A (dBA) | ISO/ANSI (dBA) | Difference (ISO/ANSI - 866A) (dBA) |
|---|---|---|---|
| 200 | 103.9 | 103.9 | 0.0 |
| 400 | 97.7 | 97.8 | 0.0 |
| 630 | 93.6 | 93.6 | 0.1 |
| 1000 | 89.3 | 89.4 | 0.1 |
| 2000 | 82.5 | 82.7 | 0.3 |
| 4000 | 75.0 | 75.6 | 0.6 |
| 6300 | 69.5 | 70.5 | 1.0 |
| 10000 | 63.1 | 64.8 | 1.8 |
| 16000 | 55.5 | 58.5 | 3.0 |
| 25000 | 46.9 | 51.9 | 4.9 |

**Table C.61** Spectral Class 302 at 85°F/85% RH.

| Distance (ft) | 866A (dBA) | ISO/ANSI (dBA) | Difference (ISO/ANSI - 866A) (dBA) |
|---|---|---|---|
| 200 | 93.6 | 93.4 | -0.2 |
| 400 | 87.0 | 86.8 | -0.2 |
| 630 | 82.4 | 82.2 | -0.2 |
| 1000 | 77.4 | 77.2 | -0.2 |
| 2000 | 69.1 | 68.8 | -0.3 |
| 4000 | 59.5 | 59.3 | -0.2 |
| 6300 | 52.4 | 52.5 | 0.2 |
| 10000 | 44.6 | 45.8 | 1.2 |
| 16000 | 36.3 | 39.0 | 2.7 |
| 25000 | 27.9 | 32.6 | 4.7 |

**Table C.62** Spectral Class 307 at 85°F/85% RH.

| Distance (ft) | 866A (dBA) | ISO/ANSI (dBA) | Difference (ISO/ANSI - 866A) (dBA) |
|---|---|---|---|
| 200 | 97.3 | 97.3 | -0.1 |
| 400 | 91.0 | 90.9 | 0.0 |
| 630 | 86.7 | 86.7 | 0.0 |
| 1000 | 82.2 | 82.2 | 0.0 |
| 2000 | 75.0 | 75.1 | 0.1 |
| 4000 | 67.3 | 67.7 | 0.5 |
| 6300 | 61.7 | 62.6 | 0.9 |
| 10000 | 55.5 | 57.2 | 1.7 |
| 16000 | 48.5 | 51.5 | 3.0 |
| 25000 | 40.9 | 45.7 | 4.8 |

**Table C.63** MD900 Spectral Data at 85°F/85% RH.

| Distance (ft) | 866A (dBA) | ISO/ANSI (dBA) | Difference (ISO/ANSI - 866A) (dBA) |
|---|---|---|---|
| 200 | 81.2 | 81.2 | 0.0 |
| 400 | 74.8 | 74.7 | 0.0 |
| 630 | 70.4 | 70.4 | 0.0 |
| 1000 | 65.8 | 65.7 | 0.0 |
| 2000 | 58.3 | 58.3 | 0.0 |
| 4000 | 50.0 | 50.3 | 0.3 |
| 6300 | 44.0 | 44.7 | 0.7 |
| 10000 | 37.2 | 38.7 | 1.6 |
| 16000 | 29.4 | 32.3 | 2.9 |
| 25000 | 20.9 | 25.7 | 4.8 |

## C.4 As-Measured Spectra Adjusted to 40°F/85% RH

**Table C.64** Spectral Class 101 at 40°F/85% RH.

| Distance (ft) | 866A (dBA) | ISO/ANSI (dBA) | Difference (ISO/ANSI - 866A) (dBA) |
|---|---|---|---|
| 200 | 94.9 | 94.9 | 0.0 |
| 400 | 88.6 | 88.6 | 0.0 |
| 630 | 84.3 | 84.4 | 0.0 |
| 1000 | 79.8 | 80.0 | 0.1 |
| 2000 | 72.8 | 73.1 | 0.3 |
| 4000 | 65.1 | 65.6 | 0.5 |
| 6300 | 59.7 | 60.3 | 0.6 |
| 10000 | 53.6 | 54.4 | 0.8 |
| 16000 | 47.0 | 47.9 | 0.9 |
| 25000 | 40.1 | 41.1 | 1.0 |

**Table C.65** Spectral Class 102 at 40°F/85% RH.

| Distance (ft) | 866A (dBA) | ISO/ANSI (dBA) | Difference (ISO/ANSI - 866A) (dBA) |
|---|---|---|---|
| 200 | 94.7 | 94.7 | 0.0 |
| 400 | 88.3 | 88.3 | 0.0 |
| 630 | 84.0 | 84.0 | 0.1 |
| 1000 | 79.4 | 79.6 | 0.2 |
| 2000 | 72.2 | 72.6 | 0.4 |
| 4000 | 64.3 | 64.9 | 0.6 |
| 6300 | 58.4 | 59.3 | 0.9 |
| 10000 | 51.8 | 53.0 | 1.2 |
| 16000 | 44.2 | 45.7 | 1.5 |
| 25000 | 36.1 | 37.9 | 1.8 |

**Table C.66** Spectral Class 103 at 40°F/85% RH.

| Distance (ft) | 866A (dBA) | ISO/ANSI (dBA) | Difference (ISO/ANSI - 866A) (dBA) |
|---|---|---|---|
| 200 | 96.7 | 96.4 | -0.2 |
| 400 | 90.1 | 89.9 | -0.2 |
| 630 | 85.6 | 85.5 | -0.1 |
| 1000 | 80.8 | 80.8 | 0.0 |
| 2000 | 73.2 | 73.4 | 0.2 |
| 4000 | 64.9 | 65.4 | 0.5 |
| 6300 | 59.2 | 59.8 | 0.7 |
| 10000 | 53.0 | 53.8 | 0.8 |
| 16000 | 46.2 | 47.1 | 0.9 |
| 25000 | 39.0 | 40.1 | 1.0 |

**Table C.67** Spectral Class 104 at 40°F/85% RH.

| Distance (ft) | 866A (dBA) | ISO/ANSI (dBA) | Difference (ISO/ANSI - 866A) (dBA) |
|---|---|---|---|
| 200 | 94.0 | 94.0 | 0.0 |
| 400 | 87.7 | 87.7 | 0.0 |
| 630 | 83.4 | 83.4 | 0.1 |
| 1000 | 78.9 | 79.0 | 0.1 |
| 2000 | 71.7 | 72.0 | 0.3 |
| 4000 | 63.8 | 64.3 | 0.6 |
| 6300 | 58.1 | 58.8 | 0.8 |
| 10000 | 51.7 | 52.7 | 1.0 |
| 16000 | 44.5 | 45.7 | 1.2 |
| 25000 | 36.8 | 38.2 | 1.4 |

**Table C.68** Spectral Class 105 at 40°F/85% RH.

| Distance (ft) | 866A (dBA) | ISO/ANSI (dBA) | Difference (ISO/ANSI - 866A) (dBA) |
|---|---|---|---|
| 200 | 95.8 | 95.7 | -0.1 |
| 400 | 89.3 | 89.2 | -0.1 |
| 630 | 85.0 | 84.9 | 0.0 |
| 1000 | 80.3 | 80.4 | 0.1 |
| 2000 | 73.1 | 73.3 | 0.2 |
| 4000 | 65.2 | 65.7 | 0.5 |
| 6300 | 59.6 | 60.3 | 0.6 |
| 10000 | 53.6 | 54.4 | 0.8 |
| 16000 | 46.9 | 47.8 | 0.9 |
| 25000 | 39.9 | 41.0 | 1.0 |

**Table C.69** Spectral Class 106 at 40°F/85% RH.

| Distance (ft) | 866A (dBA) | ISO/ANSI (dBA) | Difference (ISO/ANSI - 866A) (dBA) |
|---|---|---|---|
| 200 | 96.2 | 96.1 | -0.1 |
| 400 | 89.7 | 89.6 | -0.1 |
| 630 | 85.3 | 85.3 | 0.0 |
| 1000 | 80.7 | 80.8 | 0.1 |
| 2000 | 73.5 | 73.8 | 0.3 |
| 4000 | 65.7 | 66.2 | 0.5 |
| 6300 | 60.1 | 60.8 | 0.7 |
| 10000 | 53.8 | 54.7 | 1.0 |
| 16000 | 46.5 | 47.8 | 1.3 |
| 25000 | 38.7 | 40.2 | 1.6 |

**Table C.70** Spectral Class 107 at 40°F/85% RH.

| Distance (ft) | 866A (dBA) | ISO/ANSI (dBA) | Difference (ISO/ANSI - 866A) (dBA) |
|---|---|---|---|
| 200 | 97.9 | 97.5 | -0.3 |
| 400 | 91.2 | 90.9 | -0.3 |
| 630 | 86.5 | 86.3 | -0.2 |
| 1000 | 81.5 | 81.4 | -0.1 |
| 2000 | 73.5 | 73.7 | 0.2 |
| 4000 | 65.0 | 65.6 | 0.5 |
| 6300 | 59.2 | 60.0 | 0.7 |
| 10000 | 52.9 | 53.9 | 0.9 |
| 16000 | 45.9 | 47.0 | 1.1 |
| 25000 | 38.5 | 39.8 | 1.2 |

**Table C.71** Spectral Class 108 at 40°F/85% RH.

| Distance (ft) | 866A (dBA) | ISO/ANSI (dBA) | Difference (ISO/ANSI - 866A) (dBA) |
|---|---|---|---|
| 200 | 95.3 | 95.1 | -0.1 |
| 400 | 88.4 | 88.2 | -0.2 |
| 630 | 83.8 | 83.6 | -0.1 |
| 1000 | 78.9 | 78.9 | 0.0 |
| 2000 | 71.3 | 71.6 | 0.3 |
| 4000 | 63.2 | 63.8 | 0.6 |
| 6300 | 57.3 | 58.1 | 0.8 |
| 10000 | 51.0 | 52.0 | 1.0 |
| 16000 | 44.0 | 45.1 | 1.1 |
| 25000 | 37.0 | 38.2 | 1.2 |

**Table C.72** Spectral Class 109 at 40°F/85% RH.

| Distance (ft) | 866A (dBA) | ISO/ANSI (dBA) | Difference (ISO/ANSI - 866A) (dBA) |
|---|---|---|---|
| 200 | 100.1 | 100.2 | 0.0 |
| 400 | 94.0 | 94.0 | 0.0 |
| 630 | 89.9 | 89.9 | 0.1 |
| 1000 | 85.6 | 85.7 | 0.1 |
| 2000 | 79.0 | 79.2 | 0.2 |
| 4000 | 71.9 | 72.3 | 0.3 |
| 6300 | 66.9 | 67.3 | 0.5 |
| 10000 | 61.2 | 61.8 | 0.6 |
| 16000 | 54.9 | 55.6 | 0.8 |
| 25000 | 48.2 | 49.1 | 0.9 |

**Table C.73** Spectral Class 110 at 40°F/85% RH.

| Distance (ft) | 866A (dBA) | ISO/ANSI (dBA) | Difference (ISO/ANSI - 866A) (dBA) |
|---|---|---|---|
| 200 | 100.3 | 100.2 | -0.1 |
| 400 | 93.9 | 93.9 | -0.1 |
| 630 | 89.8 | 89.7 | 0.0 |
| 1000 | 85.4 | 85.4 | 0.0 |
| 2000 | 78.7 | 78.8 | 0.1 |
| 4000 | 71.7 | 71.9 | 0.2 |
| 6300 | 66.7 | 67.0 | 0.3 |
| 10000 | 61.3 | 61.7 | 0.4 |
| 16000 | 55.1 | 55.6 | 0.6 |
| 25000 | 48.3 | 49.1 | 0.8 |

**Table C.74** Spectral Class 111 at 40°F/85% RH.

| Distance (ft) | 866A (dBA) | ISO/ANSI (dBA) | Difference (ISO/ANSI - 866A) (dBA) |
|---|---|---|---|
| 200 | 106.2 | 106.1 | -0.1 |
| 400 | 100.1 | 100.1 | -0.1 |
| 630 | 96.1 | 96.0 | -0.1 |
| 1000 | 91.9 | 91.9 | 0.0 |
| 2000 | 85.6 | 85.6 | 0.0 |
| 4000 | 78.9 | 79.0 | 0.1 |
| 6300 | 74.3 | 74.4 | 0.1 |
| 10000 | 69.2 | 69.5 | 0.2 |
| 16000 | 63.6 | 64.0 | 0.4 |
| 25000 | 57.6 | 58.2 | 0.7 |

**Table C.75** Spectral Class 112 at 40°F/85% RH.

| Distance (ft) | 866A (dBA) | ISO/ANSI (dBA) | Difference (ISO/ANSI - 866A) (dBA) |
|---|---|---|---|
| 200 | 100.2 | 100.0 | -0.2 |
| 400 | 93.9 | 93.7 | -0.2 |
| 630 | 89.6 | 89.5 | -0.1 |
| 1000 | 85.1 | 85.1 | 0.0 |
| 2000 | 78.3 | 78.3 | 0.1 |
| 4000 | 71.2 | 71.4 | 0.2 |
| 6300 | 66.5 | 66.7 | 0.2 |
| 10000 | 61.6 | 61.9 | 0.3 |
| 16000 | 56.3 | 56.7 | 0.4 |
| 25000 | 50.9 | 51.5 | 0.6 |

**Table C.76** Spectral Class 113 at 40°F/85% RH.

| Distance (ft) | 866A (dBA) | ISO/ANSI (dBA) | Difference (ISO/ANSI - 866A) (dBA) |
|---|---|---|---|
| 200 | 94.6 | 94.6 | -0.1 |
| 400 | 88.3 | 88.2 | 0.0 |
| 630 | 83.9 | 84.0 | 0.0 |
| 1000 | 79.4 | 79.5 | 0.1 |
| 2000 | 72.1 | 72.5 | 0.3 |
| 4000 | 64.2 | 64.8 | 0.6 |
| 6300 | 58.4 | 59.2 | 0.8 |
| 10000 | 52.0 | 53.0 | 1.0 |
| 16000 | 44.7 | 46.0 | 1.3 |
| 25000 | 36.8 | 38.3 | 1.5 |

**Table C.77** Spectral Class 116 at 40°F/85% RH.

| Distance (ft) | 866A (dBA) | ISO/ANSI (dBA) | Difference (ISO/ANSI - 866A) (dBA) |
|---|---|---|---|
| 200 | 95.7 | 95.7 | 0.0 |
| 400 | 89.4 | 89.5 | 0.0 |
| 630 | 85.3 | 85.3 | 0.1 |
| 1000 | 80.9 | 81.0 | 0.1 |
| 2000 | 74.0 | 74.2 | 0.3 |
| 4000 | 66.4 | 66.9 | 0.5 |
| 6300 | 61.0 | 61.6 | 0.6 |
| 10000 | 54.9 | 55.7 | 0.9 |
| 16000 | 47.9 | 49.0 | 1.1 |
| 25000 | 40.3 | 41.6 | 1.3 |

**Table C.78** Spectral Class 203 at 40°F/85% RH.

| Distance (ft) | 866A (dBA) | ISO/ANSI (dBA) | Difference (ISO/ANSI - 866A) (dBA) |
|---|---|---|---|
| 200 | 98.5 | 98.1 | -0.5 |
| 400 | 91.5 | 91.0 | -0.5 |
| 630 | 86.6 | 86.2 | -0.4 |
| 1000 | 81.3 | 81.1 | -0.2 |
| 2000 | 73.2 | 73.4 | 0.2 |
| 4000 | 64.8 | 65.4 | 0.6 |
| 6300 | 59.0 | 59.8 | 0.8 |
| 10000 | 52.6 | 53.6 | 1.0 |
| 16000 | 45.3 | 46.5 | 1.2 |
| 25000 | 37.6 | 39.0 | 1.4 |

**Table C.79** Spectral Class 205 at 40°F/85% RH.

| Distance (ft) | 866A (dBA) | ISO/ANSI (dBA) | Difference (ISO/ANSI - 866A) (dBA) |
|---|---|---|---|
| 200 | 95.2 | 95.1 | -0.1 |
| 400 | 88.6 | 88.5 | -0.1 |
| 630 | 84.2 | 84.1 | -0.1 |
| 1000 | 79.4 | 79.5 | 0.1 |
| 2000 | 72.0 | 72.3 | 0.3 |
| 4000 | 64.0 | 64.6 | 0.6 |
| 6300 | 58.3 | 59.0 | 0.8 |
| 10000 | 51.9 | 52.9 | 1.0 |
| 16000 | 44.8 | 46.0 | 1.2 |
| 25000 | 37.6 | 38.8 | 1.3 |

**Table C.80** Spectral Class 219 at 40°F/85% RH.

| Distance (ft) | 866A (dBA) | ISO/ANSI (dBA) | Difference (ISO/ANSI - 866A) (dBA) |
|---|---|---|---|
| 200 | 97.6 | 97.6 | 0.0 |
| 400 | 91.5 | 91.5 | 0.0 |
| 630 | 87.3 | 87.4 | 0.0 |
| 1000 | 83.0 | 83.1 | 0.1 |
| 2000 | 76.4 | 76.5 | 0.2 |
| 4000 | 69.2 | 69.5 | 0.3 |
| 6300 | 64.2 | 64.6 | 0.4 |
| 10000 | 58.7 | 59.1 | 0.5 |
| 16000 | 52.4 | 53.1 | 0.6 |
| 25000 | 45.8 | 46.6 | 0.8 |

**Table C.81** Spectral Class 222 at 40°F/85% RH.

| Distance (ft) | 866A (dBA) | ISO/ANSI (dBA) | Difference (ISO/ANSI - 866A) (dBA) |
|---|---|---|---|
| 200 | 104.0 | 104.0 | 0.0 |
| 400 | 97.8 | 97.8 | 0.0 |
| 630 | 93.8 | 93.8 | 0.0 |
| 1000 | 89.6 | 89.6 | 0.0 |
| 2000 | 83.0 | 83.1 | 0.1 |
| 4000 | 76.1 | 76.3 | 0.2 |
| 6300 | 71.1 | 71.4 | 0.3 |
| 10000 | 65.5 | 66.0 | 0.5 |
| 16000 | 59.1 | 59.8 | 0.7 |
| 25000 | 51.9 | 52.9 | 1.0 |

**Table C.82** Spectral Class 302 at 40°F/85% RH.

| Distance (ft) | 866A (dBA) | ISO/ANSI (dBA) | Difference (ISO/ANSI - 866A) (dBA) |
|---|---|---|---|
| 200 | 93.7 | 93.5 | -0.1 |
| 400 | 87.1 | 87.0 | -0.1 |
| 630 | 82.7 | 82.6 | 0.0 |
| 1000 | 77.9 | 78.0 | 0.1 |
| 2000 | 70.2 | 70.5 | 0.3 |
| 4000 | 61.5 | 62.2 | 0.7 |
| 6300 | 55.1 | 56.1 | 1.0 |
| 10000 | 48.1 | 49.4 | 1.3 |
| 16000 | 40.5 | 41.8 | 1.4 |
| 25000 | 32.7 | 34.2 | 1.4 |

**Table C.83** Spectral Class 307 at 40°F/85% RH.

| Distance (ft) | 866A (dBA) | ISO/ANSI (dBA) | Difference (ISO/ANSI - 866A) (dBA) |
|---|---|---|---|
| 200 | 97.4 | 97.3 | -0.1 |
| 400 | 91.1 | 91.1 | 0.0 |
| 630 | 86.9 | 86.9 | 0.0 |
| 1000 | 82.5 | 82.5 | 0.1 |
| 2000 | 75.7 | 75.8 | 0.2 |
| 4000 | 68.4 | 68.7 | 0.3 |
| 6300 | 63.3 | 63.7 | 0.4 |
| 10000 | 57.7 | 58.2 | 0.5 |
| 16000 | 51.5 | 52.1 | 0.7 |
| 25000 | 44.9 | 45.7 | 0.8 |

**Table C.84** MD900 Spectral Data at 40°F/85% RH.

| Distance (ft) | 866A (dBA) | ISO/ANSI (dBA) | Difference (ISO/ANSI - 866A) (dBA) |
|---|---|---|---|
| 200 | 81.3 | 81.2 | -0.1 |
| 400 | 74.9 | 74.9 | 0.0 |
| 630 | 70.6 | 70.6 | 0.0 |
| 1000 | 66.1 | 66.2 | 0.1 |
| 2000 | 59.1 | 59.3 | 0.3 |
| 4000 | 51.5 | 51.9 | 0.5 |
| 6300 | 46.0 | 46.6 | 0.6 |
| 10000 | 40.0 | 40.7 | 0.8 |
| 16000 | 33.1 | 34.1 | 1.0 |
| 25000 | 25.8 | 27.0 | 1.2 |

## C.5 As-Measured Spectra Adjusted to 40°F/55% RH

**Table C.85** Spectral Class 101 at 40°F/55% RH.

| Distance (ft) | 866A (dBA) | ISO/ANSI (dBA) | Difference (ISO/ANSI - 866A) (dBA) |
|---|---|---|---|
| 200 | 94.8 | 94.7 | 0.0 |
| 400 | 88.3 | 88.4 | 0.0 |
| 630 | 84.0 | 84.1 | 0.1 |
| 1000 | 79.4 | 79.6 | 0.2 |
| 2000 | 72.2 | 72.6 | 0.4 |
| 4000 | 64.4 | 65.0 | 0.6 |
| 6300 | 58.9 | 59.6 | 0.7 |
| 10000 | 53.0 | 53.8 | 0.8 |
| 16000 | 46.5 | 47.3 | 0.8 |
| 25000 | 39.8 | 40.6 | 0.8 |

**Table C.86** Spectral Class 102 at 40°F/55% RH.

| Distance (ft) | 866A (dBA) | ISO/ANSI (dBA) | Difference (ISO/ANSI - 866A) (dBA) |
|---|---|---|---|
| 200 | 94.5 | 94.5 | 0.0 |
| 400 | 88.0 | 88.0 | 0.1 |
| 630 | 83.6 | 83.7 | 0.1 |
| 1000 | 78.9 | 79.2 | 0.3 |
| 2000 | 71.5 | 72.0 | 0.5 |
| 4000 | 63.3 | 64.1 | 0.8 |
| 6300 | 57.3 | 58.4 | 1.1 |
| 10000 | 50.7 | 52.0 | 1.3 |
| 16000 | 43.1 | 44.7 | 1.6 |
| 25000 | 35.3 | 36.9 | 1.6 |

**Table C.87** Spectral Class 103 at 40°F/55% RH.

| Distance (ft) | 866A (dBA) | ISO/ANSI (dBA) | Difference (ISO/ANSI - 866A) (dBA) |
|---|---|---|---|
| 200 | 96.4 | 96.2 | -0.2 |
| 400 | 89.6 | 89.5 | -0.1 |
| 630 | 84.9 | 84.9 | 0.0 |
| 1000 | 80.0 | 80.1 | 0.1 |
| 2000 | 72.1 | 72.5 | 0.4 |
| 4000 | 63.9 | 64.5 | 0.6 |
| 6300 | 58.3 | 59.0 | 0.7 |
| 10000 | 52.3 | 53.1 | 0.8 |
| 16000 | 45.7 | 46.6 | 0.9 |
| 25000 | 38.7 | 39.7 | 0.9 |

**Table C.88** Spectral Class 104 at 40°F/55% RH.

| Distance (ft) | 866A (dBA) | ISO/ANSI (dBA) | Difference (ISO/ANSI - 866A) (dBA) |
|---|---|---|---|
| 200 | 93.9 | 93.8 | 0.0 |
| 400 | 87.4 | 87.5 | 0.1 |
| 630 | 83.0 | 83.1 | 0.1 |
| 1000 | 78.4 | 78.6 | 0.2 |
| 2000 | 71.0 | 71.4 | 0.4 |
| 4000 | 62.9 | 63.6 | 0.7 |
| 6300 | 57.1 | 58.0 | 0.9 |
| 10000 | 50.8 | 51.8 | 1.0 |
| 16000 | 43.8 | 45.0 | 1.2 |
| 25000 | 36.3 | 37.6 | 1.4 |

**Table C.89** Spectral Class 105 at 40°F/55% RH.

| Distance (ft) | 866A (dBA) | ISO/ANSI (dBA) | Difference (ISO/ANSI - 866A) (dBA) |
|---|---|---|---|
| 200 | 95.6 | 95.5 | -0.1 |
| 400 | 89.0 | 88.9 | 0.0 |
| 630 | 84.5 | 84.5 | 0.0 |
| 1000 | 79.7 | 79.9 | 0.2 |
| 2000 | 72.3 | 72.7 | 0.4 |
| 4000 | 64.4 | 65.0 | 0.6 |
| 6300 | 58.9 | 59.6 | 0.7 |
| 10000 | 53.0 | 53.7 | 0.8 |
| 16000 | 46.4 | 47.3 | 0.8 |
| 25000 | 39.6 | 40.5 | 0.9 |

**Table C.90** Spectral Class 106 at 40°F/55% RH.

| Distance (ft) | 866A (dBA) | ISO/ANSI (dBA) | Difference (ISO/ANSI - 866A) (dBA) |
|---|---|---|---|
| 200 | 96.0 | 95.9 | -0.1 |
| 400 | 89.4 | 89.3 | 0.0 |
| 630 | 84.9 | 85.0 | 0.1 |
| 1000 | 80.2 | 80.4 | 0.2 |
| 2000 | 72.9 | 73.3 | 0.4 |
| 4000 | 64.9 | 65.6 | 0.7 |
| 6300 | 59.3 | 60.1 | 0.9 |
| 10000 | 52.9 | 54.0 | 1.1 |
| 16000 | 45.7 | 47.1 | 1.4 |
| 25000 | 38.0 | 39.5 | 1.5 |

**Table C.91** Spectral Class 107 at 40°F/55% RH.

| Distance (ft) | 866A (dBA) | ISO/ANSI (dBA) | Difference (ISO/ANSI - 866A) (dBA) |
|---|---|---|---|
| 200 | 97.5 | 97.2 | -0.3 |
| 400 | 90.5 | 90.3 | -0.2 |
| 630 | 85.7 | 85.5 | -0.1 |
| 1000 | 80.4 | 80.5 | 0.1 |
| 2000 | 72.3 | 72.7 | 0.4 |
| 4000 | 64.0 | 64.7 | 0.7 |
| 6300 | 58.3 | 59.1 | 0.8 |
| 10000 | 52.1 | 53.1 | 1.0 |
| 16000 | 45.3 | 46.3 | 1.1 |
| 25000 | 38.1 | 39.2 | 1.1 |

**Table C.92** Spectral Class 108 at 40°F/55% RH.

| Distance (ft) | 866A (dBA) | ISO/ANSI (dBA) | Difference (ISO/ANSI - 866A) (dBA) |
|---|---|---|---|
| 200 | 94.8 | 94.7 | -0.2 |
| 400 | 87.8 | 87.7 | -0.1 |
| 630 | 83.1 | 83.0 | 0.0 |
| 1000 | 78.1 | 78.3 | 0.2 |
| 2000 | 70.4 | 70.9 | 0.5 |
| 4000 | 62.1 | 62.8 | 0.7 |
| 6300 | 56.3 | 57.2 | 0.9 |
| 10000 | 50.0 | 51.0 | 1.0 |
| 16000 | 43.4 | 44.3 | 1.0 |
| 25000 | 36.7 | 37.5 | 0.9 |

**Table C.93** Spectral Class 109 at 40°F/55% RH.

| Distance (ft) | 866A (dBA) | ISO/ANSI (dBA) | Difference (ISO/ANSI - 866A) (dBA) |
|---|---|---|---|
| 200 | 100.1 | 100.1 | 0.0 |
| 400 | 93.9 | 94.0 | 0.1 |
| 630 | 89.8 | 89.9 | 0.1 |
| 1000 | 85.5 | 85.6 | 0.1 |
| 2000 | 78.8 | 79.1 | 0.2 |
| 4000 | 71.6 | 72.0 | 0.4 |
| 6300 | 66.5 | 67.0 | 0.5 |
| 10000 | 60.8 | 61.4 | 0.6 |
| 16000 | 54.5 | 55.2 | 0.7 |
| 25000 | 48.0 | 48.7 | 0.7 |

**Table C.94** Spectral Class 110 at 40°F/55% RH.

| Distance (ft) | 866A (dBA) | ISO/ANSI (dBA) | Difference (ISO/ANSI - 866A) (dBA) |
|---|---|---|---|
| 200 | 100.1 | 100.1 | -0.1 |
| 400 | 93.8 | 93.7 | 0.0 |
| 630 | 89.6 | 89.5 | 0.0 |
| 1000 | 85.2 | 85.2 | 0.0 |
| 2000 | 78.5 | 78.6 | 0.1 |
| 4000 | 71.4 | 71.7 | 0.2 |
| 6300 | 66.5 | 66.8 | 0.3 |
| 10000 | 61.1 | 61.5 | 0.4 |
| 16000 | 54.9 | 55.5 | 0.6 |
| 25000 | 48.2 | 48.9 | 0.7 |

**Table C.95** Spectral Class 111 at 40°F/55% RH.

| Distance (ft) | 866A (dBA) | ISO/ANSI (dBA) | Difference (ISO/ANSI - 866A) (dBA) |
|---|---|---|---|
| 200 | 106.2 | 106.1 | -0.1 |
| 400 | 100.1 | 100.0 | -0.1 |
| 630 | 96.1 | 96.0 | -0.1 |
| 1000 | 91.9 | 91.9 | 0.0 |
| 2000 | 85.5 | 85.5 | 0.0 |
| 4000 | 78.9 | 78.9 | 0.1 |
| 6300 | 74.2 | 74.3 | 0.1 |
| 10000 | 69.2 | 69.4 | 0.2 |
| 16000 | 63.5 | 63.8 | 0.3 |
| 25000 | 57.5 | 58.0 | 0.5 |

**Table C.96** Spectral Class 112 at 40°F/55% RH.

| Distance (ft) | 866A (dBA) | ISO/ANSI (dBA) | Difference (ISO/ANSI - 866A) (dBA) |
|---|---|---|---|
| 200 | 100.0 | 99.9 | -0.2 |
| 400 | 93.6 | 93.4 | -0.1 |
| 630 | 89.2 | 89.1 | -0.1 |
| 1000 | 84.7 | 84.7 | 0.0 |
| 2000 | 77.9 | 78.0 | 0.1 |
| 4000 | 71.0 | 71.1 | 0.1 |
| 6300 | 66.3 | 66.5 | 0.1 |
| 10000 | 61.5 | 61.6 | 0.1 |
| 16000 | 56.2 | 56.4 | 0.2 |
| 25000 | 50.9 | 51.1 | 0.2 |

**Table C.97** Spectral Class 113 at 40°F/55% RH.

| Distance (ft) | 866A (dBA) | ISO/ANSI (dBA) | Difference (ISO/ANSI - 866A) (dBA) |
|---|---|---|---|
| 200 | 94.5 | 94.4 | 0.0 |
| 400 | 88.0 | 88.0 | 0.0 |
| 630 | 83.5 | 83.6 | 0.1 |
| 1000 | 78.8 | 79.1 | 0.2 |
| 2000 | 71.4 | 71.8 | 0.4 |
| 4000 | 63.3 | 64.0 | 0.7 |
| 6300 | 57.5 | 58.4 | 0.9 |
| 10000 | 51.1 | 52.2 | 1.1 |
| 16000 | 43.9 | 45.2 | 1.3 |
| 25000 | 36.2 | 37.7 | 1.4 |

**Table C.98** Spectral Class 116 at 40°F/55% RH.

| Distance (ft) | 866A (dBA) | ISO/ANSI (dBA) | Difference (ISO/ANSI - 866A) (dBA) |
|---|---|---|---|
| 200 | 95.6 | 95.6 | 0.0 |
| 400 | 89.3 | 89.4 | 0.1 |
| 630 | 85.0 | 85.1 | 0.1 |
| 1000 | 80.6 | 80.8 | 0.2 |
| 2000 | 73.5 | 73.9 | 0.4 |
| 4000 | 65.9 | 66.4 | 0.6 |
| 6300 | 60.3 | 61.1 | 0.8 |
| 10000 | 54.2 | 55.1 | 1.0 |
| 16000 | 47.2 | 48.4 | 1.2 |
| 25000 | 39.8 | 41.1 | 1.3 |

**Table C.99** Spectral Class 203 at 40°F/55% RH.

| Distance (ft) | 866A (dBA) | ISO/ANSI (dBA) | Difference (ISO/ANSI - 866A) (dBA) |
|---|---|---|---|
| 200 | 98.0 | 97.5 | -0.5 |
| 400 | 90.6 | 90.2 | -0.4 |
| 630 | 85.5 | 85.2 | -0.2 |
| 1000 | 80.1 | 80.1 | 0.0 |
| 2000 | 72.1 | 72.5 | 0.4 |
| 4000 | 63.8 | 64.5 | 0.7 |
| 6300 | 58.0 | 58.9 | 0.9 |
| 10000 | 51.6 | 52.7 | 1.1 |
| 16000 | 44.5 | 45.8 | 1.2 |
| 25000 | 37.1 | 38.4 | 1.3 |

**Table C.100** Spectral Class 205 at 40°F/55% RH.

| Distance (ft) | 866A (dBA) | ISO/ANSI (dBA) | Difference (ISO/ANSI - 866A) (dBA) |
|---|---|---|---|
| 200 | 94.9 | 94.8 | -0.1 |
| 400 | 88.2 | 88.1 | -0.1 |
| 630 | 83.6 | 83.6 | 0.0 |
| 1000 | 78.8 | 79.0 | 0.2 |
| 2000 | 71.2 | 71.6 | 0.4 |
| 4000 | 63.1 | 63.8 | 0.7 |
| 6300 | 57.3 | 58.2 | 0.9 |
| 10000 | 51.0 | 52.0 | 1.0 |
| 16000 | 44.2 | 45.2 | 1.1 |
| 25000 | 37.1 | 38.2 | 1.0 |

**Table C.101** Spectral Class 219 at 40°F/55% RH.

| Distance (ft) | 866A (dBA) | ISO/ANSI (dBA) | Difference (ISO/ANSI - 866A) (dBA) |
|---|---|---|---|
| 200 | 97.6 | 97.6 | 0.0 |
| 400 | 91.4 | 91.4 | 0.0 |
| 630 | 87.2 | 87.2 | 0.1 |
| 1000 | 82.9 | 82.9 | 0.1 |
| 2000 | 76.1 | 76.3 | 0.2 |
| 4000 | 68.9 | 69.2 | 0.3 |
| 6300 | 63.9 | 64.3 | 0.4 |
| 10000 | 58.4 | 58.8 | 0.5 |
| 16000 | 52.2 | 52.8 | 0.6 |
| 25000 | 45.7 | 46.3 | 0.7 |

**Table C.102** Spectral Class 222 at 40°F/55% RH.

| Distance (ft) | 866A (dBA) | ISO/ANSI (dBA) | Difference (ISO/ANSI - 866A) (dBA) |
|---|---|---|---|
| 200 | 104.0 | 103.9 | 0.0 |
| 400 | 97.8 | 97.8 | 0.0 |
| 630 | 93.7 | 93.7 | 0.0 |
| 1000 | 89.5 | 89.6 | 0.1 |
| 2000 | 82.9 | 83.1 | 0.1 |
| 4000 | 75.9 | 76.2 | 0.3 |
| 6300 | 70.9 | 71.3 | 0.4 |
| 10000 | 65.3 | 65.8 | 0.6 |
| 16000 | 58.8 | 59.6 | 0.8 |
| 25000 | 51.7 | 52.7 | 1.0 |

**Table C.103** Spectral Class 302 at 40°F/55% RH.

| Distance (ft) | 866A (dBA) | ISO/ANSI (dBA) | Difference (ISO/ANSI - 866A) (dBA) |
|---|---|---|---|
| 200 | 93.4 | 93.3 | -0.1 |
| 400 | 86.7 | 86.6 | 0.0 |
| 630 | 82.0 | 82.1 | 0.1 |
| 1000 | 77.0 | 77.2 | 0.2 |
| 2000 | 68.9 | 69.5 | 0.6 |
| 4000 | 59.9 | 60.8 | 0.9 |
| 6300 | 53.5 | 54.6 | 1.1 |
| 10000 | 46.7 | 48.0 | 1.2 |
| 16000 | 39.5 | 40.8 | 1.2 |
| 25000 | 32.2 | 33.4 | 1.2 |

**Table C.104** Spectral Class 307 at 40°F/55% RH.

| Distance (ft) | 866A (dBA) | ISO/ANSI (dBA) | Difference (ISO/ANSI - 866A) (dBA) |
|---|---|---|---|
| 200 | 97.3 | 97.2 | -0.1 |
| 400 | 90.9 | 90.9 | 0.0 |
| 630 | 86.6 | 86.7 | 0.0 |
| 1000 | 82.2 | 82.3 | 0.1 |
| 2000 | 75.3 | 75.5 | 0.2 |
| 4000 | 68.0 | 68.4 | 0.3 |
| 6300 | 62.9 | 63.3 | 0.4 |
| 10000 | 57.4 | 57.9 | 0.5 |
| 16000 | 51.2 | 51.8 | 0.6 |
| 25000 | 44.7 | 45.4 | 0.7 |

**Table C.105** MD900 Spectral Data at 40°F/55% RH.

| Distance (ft) | 866A (dBA) | ISO/ANSI (dBA) | Difference (ISO/ANSI - 866A) (dBA) |
|---|---|---|---|
| 200 | 81.1 | 81.1 | 0.0 |
| 400 | 74.7 | 74.7 | 0.0 |
| 630 | 70.3 | 70.4 | 0.1 |
| 1000 | 65.7 | 65.9 | 0.2 |
| 2000 | 58.5 | 58.9 | 0.3 |
| 4000 | 50.8 | 51.3 | 0.5 |
| 6300 | 45.4 | 46.0 | 0.7 |
| 10000 | 39.4 | 40.2 | 0.8 |
| 16000 | 32.7 | 33.6 | 1.0 |
| 25000 | 25.5 | 26.6 | 1.1 |